# The Female Turn

Malin Ah-King

# The Female Turn

## How Evolutionary Science Shifted Perceptions About Females

palgrave
macmillan

Malin Ah-King
Department of Ethnology, History of Religions and Gender Studies
Stockholm University
Stockholm, Sweden

ISBN 978-981-19-7160-0 ISBN 978-981-19-7161-7 (eBook)
https://doi.org/10.1007/978-981-19-7161-7



Cover illustration: Sabena Jane Blackbird / Alamy Stock Photo

This Palgrave Macmillan imprint is published by the registered company Springer Nature Singapore Pte Ltd.
The registered company address is: 152 Beach Road, #21-01/04 Gateway East, Singapore 189721, Singapore

*To all those struggling before, alongside and subsequent to me*

# Acknowledgments

I acknowledge the Sámi people as the indigenous people of my country, Sweden, and that the Swedish state has and continues to pursue colonial power over Sápmi.

This project was funded by the Swedish Research Council under grant number 2016-02048. I have also received a travel grant from Helge Ax:son Johnson Stiftelse and a one-week research stay at Drakamöllan Nordiskt Forum för Kultur och Vetenskap.

Figures 6.2 and 7.1 are reprinted from Ah-King, M. "The history of sexual selection research provides insights as to why females are still understudied" Nature Communications 13, 6976 (2022) under Creative Commons CC-BY 4.0 license. Likewise, the discussion about male precedence in research on male trait evolution (in chapter 3), in interpretations of female multiple mating in birds (in chapter 4), in post-copulatory sexual selection, genital evolution research and sexual conflict research (in chapter 6) as well as the discussion about male precedence and the prevailing definition of sexual selection (in chapter 7) are revised from the same paper.

My deepest thanks to the researchers participating in interviews and answering questions via e-mail—for your time and effort—you have brought invaluable contributions to the study.

Thanks to Joshua Pitt, my editor at Palgrave Macmillan, for believing in my project and securing the book contract!

Thanks to the three anonymous reviewers of the book manuscript—your constructive, fair criticism and encouraging guidance were immensely helpful and I am deeply grateful!

I also want to thank my colleagues in gender studies at Stockholm University for their constructive support, giving comments on my many drafts, and encouragement during the course of the project. Furthermore, Erika Milam inspired me to take a historical perspective on sexual selection and interviewing researchers, Isabelle Dussauge gave crucial input at a pivotal moment in the application writing, and so did Staffan Bergwik.

Mikael Mildén, J.P. Drury, Sören Nylin, Karen Warkentin, Ingrid Ahnesjö, Patricia Gowaty, Linda Fuselier, Caitlin McDonough, Henning Brüllhoff, Lisa Käll and Åsa Eriksson for comments on earlier drafts. A special thanks to Sofie Tornhill for superb writing advice and Cecilia Åse for book proposal reading!

Anna Bohlin, Helena Tinnerholm-Ljungberg and Åsa Eriksson for sharing joys, fears and mutual support in our precarious lives in academia. I would not have managed through the pandemic without your company!

Jennifer, Stephen and Izana, thanks for hosting and guiding us in the nature and culture of Mauritius, you made us more Mauritian!

Micke, my emergency editor, fellow nature explorer, invaluable support and my *somebody!*

L & L—thanks for your support and all the joys and troubles!

# Contents

# List of Figures

# 1

# Introduction

Science has historically depicted female animals, including humans, as coy, passive, elusive, noncompetitive and sexually reserved,[1] in line with societal ideologies about women. In 1871, when the founder of evolutionary biology, Charles Darwin, presented his theory of sexual selection (the struggle for mates), a theory to explain evolution of sex differences, he described females as generally coy and passive in sexual encounters.[2] Males, on the other hand, he depicted as active and competing for females through combat among themselves and by charming females with their appearances and songs. Yet animal females choose male mates (but in a rather passive and reluctant way), Darwin was steadfast on this point until his death.[3] According to Darwin, they were the less eager sex, "coy, and may often be seen endeavouring for a long time to escape from the male".[4] Furthermore, a female accepts not the male that attracts her the most, but "the one which is the least distasteful".[5] These descriptions have formed perceptions about females in evolutionary biology for a long time.

[1] Hrdy, 2000: xiv.

[2] Darwin 1871, part I: 272-273.

[3] Richards, 2017: 517.

[4] Darwin 1871 part I: 273.

[5] Darwin 1871 part I: 273.



M. Ah-King, *The Female Turn*, https://doi.org/10.1007/978-981-19-7161-7_1

Since then, evolutionary biology has changed and continues to do so. Classical perceptions of females as "coy" and passive in sexual encounters have bit by bit and at least partly shifted to acknowledging that females can have active sexual strategies, and initiate mating. Research has shown that females can also be dominant to males and fiercely competitive. Furthermore, the assumption that females are generally monogamous has been overturned by the revelation that females often mate with several males, even in most species formerly thought to be strictly monogamous. Scientific publications today encompass competitive and sexually enthusiastic females mating with multiple males. I call this still ongoing process of shifting perceptions a *female turn*.

This book investigates how and why this female turn happened (and continues), based on interviews and correspondence with experts in the international sexual selection community and the scientific record of sexual selection. Focusing on perceptions of females, I chart this particular angle of the history of sexual selection. The result is a historiography of progress in science, but this progress has not arrived momentarily nor without struggle. Traditional perceptions were criticized and defended, new suggestions met with skepticism, hence this book is a history of controversies and negotiations of scientific knowledge. Throughout the book, I examine the development of sexual selection science as a cultural and social process. Who were the pioneers? What were their contexts and influences? What resistance did they meet? How were assumptions of sexual selection challenged, defended and expanded to encompass new perceptions of females? How did these perceptions shift (or not shift) in the different subfields focusing on monkeys and apes, birds, insects, snakes, frogs, lizards, fishes and spiders?

## What is Sexual Selection?

Evolutionary theory can be applied to the question of why females and males are different, or rather—why females and males have evolved to be different. Charles Darwin's 1859 theory of natural selection provided an explanation for how species change over time, through a process in which small variations among individuals lead to differential survival or reproductive success in their environment. Certain variations that were better adapted to the current environment would therefore to a larger extent spread to the next generation. But there were features that this theory could not explain, most blatantly, the extravagant ornamental traits such as the peacock's tail. It could not possibly be explained by better survival, it should rather increase the risk of being eaten by a predator. Hence Darwin's famous quote in a letter to his friend the botanist Asa Gray in 1860:[6] "The sight of a feather in a peacock's tail, whenever I gaze at it, makes me sick!"

After years of deliberation, Darwin developed the theory of sexual selection, he suggested that these traits had evolved through "a struggle between the males for possession of the females",[7] or as he later formulated it: "Sexual selection [...] depends on the advantage which certain individuals have over other individuals of the same sex and species, in exclusive relation to reproduction".[8] If males with larger horns or spurs were more successful in gaining access to females, male-male competition could explain the evolution of armaments, and if females choose to mate with the more ornamented males, that could explain the evolution of ornaments. In *the Descent of Man and Selection in Relation to Sex,* Darwin (1871) theorized about the evolution of sex and race differences among humans and catalogued sex differences in species after species throughout the animal kingdom. He summarized that generally among animals, males are the more adorned and weaponized sex that court and compete for coy females.[9] Hence, traditionally in sexual selection, females were assumed to be "coy" and passive in sexual encounters.

Male-male competition was accepted among Darwin's contemporary biologists, but rather as part of natural selection, as they thought it could be explained by the effect of the individual's vitality and health.[10] Female mate choice in animals, however, was deeply questioned. According to the prevailing views, only humans of social upper classes had the mental

---

[6] Darwin, 1860.

[7] Darwin 1859: 103.

[8] Darwin 1871 part I: 256.

[9] Darwin 1871 part II:397.

[10] Milam, 2010: 15.

abilities for the aesthetic sense to appreciate civilization's finer aspects such as art and music—it was therefore considered highly unlikely that animals would have the complex mental abilities needed, thus female choice in animals was not acknowledged.[11] As a consequence of this resistance, sexual selection theory was overlooked, with few exceptions, until the 1970s.[12] However, science historian Erika Milam (2010) has shown that interest in sexual selection was more stretched out in time than most history writing has let us believe, for example, sexual selection was readily acknowledged by eugenicists and a popular audience as a means to improve humanity.[13]

After Darwin, later additions to sexual selection theory suggested explanations for why males had evolved to be ardent and females to be "coy" and choosy—males, but not females, gain benefits to mating with multiple partners,[14] and the typically relatively larger investment[15] in eggs and parental care makes females generally choosy and coy, while males benefit from being ardent and indiscriminate. However, since early on and throughout its development, the theory of sexual selection has been criticized from a gender perspective.

## Sexual Selection from a Gender Perspective

For hundreds of years, the sciences consistently deemed women inferior, more childlike and less intellectual than men.[16] Anatomists and craniologists at different times located the biological sex differences causing this inferiority in the skeleton, the uterus, the pelvis or the brain.[17] Then prevalent biological "facts", such as women's ovaries shrinking from too much brain use, were used to justify the exclusion of women from universities

[11] Milam, 2010:18ff, Richards, 2017:523ff.

[12] Andersson, 1994.

[13] Milam, 2010: 43ff.

[14] Bateman, 1948, but see Snyder & Gowaty, 2007, Gowaty et al., 2012.

[15] Trivers, 1972.

[16] Schiebinger, 1989, Saini, 2017.

[17] Schiebinger, 1989: 189ff.

until the late nineteenth century, and professional science even longer.[18] Evolutionary biology contributed to the construction of a hierarchical chain of beings based on sex and race differences.[19] No wonder feminists have been mistrustful and adverse toward biology. In the words of Virginia Woolf, "Science, it would seem, is not sexless; she is a man, a father, and infected too"[20]. Gender analyses of the history of science show that the exclusion of women and others than the ruling class of white men has led to gender bias and other biases in theories and research.[21] Biological research, just as medicine, has used the male as a norm[22] and thereby marginalized and hindered production of knowledge about females. In other words, it has been androcentric, that is placing the male point of view in the center of its worldview. However, ever since Charles Darwin presented his theory on sexual selection, feminists have taken issue with what it means for our understanding of sex.[23] Just four years after Darwin's *The Descent of Man and Selection in Relation to Sex* was published, American feminist Antoinette Brown Blackwell (1875) criticized Darwin for putting too much emphasis on males and ignoring females, suggesting her own theory of equality between the sexes in each species. She claimed that Darwin's and other male scientists' perspectives and theories were biased by them being men, and that women's perspectives were ignored. "I do not underrate the presumption which must attend to any woman who will attempt to controvert the great masters of science and scientific inference. There is no alternative! Only a woman can approach the subject from a feminine standpoint; and there are none but beginners among us in this class of investigations."[24] She and several other contem-

[18] Despite these obstacles some women did succeed in science during this time (Schiebinger, 1989).

[19] Schiebinger, 1989:206ff.

[20] Woolf 1938:127.

[21] Bleier, 1984:2ff, Schiebinger, 1989; see also chapter 3.

[22] Travis, 1992.

[23] Feminist philosopher Elizabeth Grosz (2011) argues for a new feminist engagement with Darwin. I agree with her intention to counteract feminist fear of essentialism and to appreciate Darwin's selection theories as describing indeterminate processes. However, I disagree with Grosz on several points, for example she describes nature as "in itself always sexed" (p. 149) which disregards variability in sex and asexual organisms. Furthermore, Grosz overlooks the development of sexual selection after Darwin.

[24] Blackwell 1875: 22.

porary feminists saw Darwin's theory as an opening of new possibilities for women in society—if male superiority over women was not instituted by God (as was generally presumed at the time), men and women could be equals.[25] But Darwin himself did not interpret sex differences in humans in terms of equality, he considered women intellectually inferior to men.[26] Darwin proposed that men, partly by natural selection and partly by sexual selection, through their constant competition among themselves, their defense of women and young from enemies, and their hunting for subsistence, had evolved higher mental qualities than women.[27] To this reasoning American feminist Caroline Kennard (1882) answered Darwin:

> "Let the 'environment' of women be similar to that of men and with his opportunities, before she be fairly judged, intellectually his inferior, please."[28]

Some years after the forceful revival of sexual selection in the 1970s, the feminist critique of androcentrism reemerged. Feminist primatologist Sarah Blaffer Hrdy in *The Woman that Never Evolved* (1981)[29] and in her essay "Empathy, polyandry[30] and the myth of the coy female" (1986)[31] criticized the ignorance of females (animal and human) in evolutionary theory and research. Recent primate studies had disclosed that females were anything but coy, questioning prevailing assumptions about female monogamy, coyness and invariability, as well as the role of males beyond mating, as carers, protectors and killers of infants. Feminist neurophysiologist Ruth Bleier's *Science and Gender: A Critique of Biology and Its Theories on Women* (1984) criticized biological determinism[32] and androcentrism in studies

---

[25] Hamlin, 2014.

[26] Darwin 1871 part II:326ff.

[27] Paradoxically, Darwin insisted that in humans, unlike in animals, men were the choosing sex (Richards, 2017: 453).

[28] Kennard, January 28, 1882. (Evans, 2017:227).

[29] Hrdy, 1981.

[30] Polyandry means that one female mates with several males.

[31] Hrdy, 1986.

[32] "Biological determinism refers to the idea that all human behavior is innate, determined by genes, brain size, or other biological attributes." Burke & Embrick, 2020.

about biological sex differences in brains, hormones and human evolution. Ornithologist Marcy Lawton and her colleagues (1997) provide an example of how gender stereotypes influenced interpretations in a behavioral study of pinyon jays, *Gymnorhinus cyanocephalus*. They analyzed ornithologists John Marzluff and Russel Balda's *The Pinyon Jay* (1992),[33] which summarized their findings from over twenty years of research. One of Marzluffs and Balda's research interests was male dominance hierarchies, but as they found pinyon jays to be very peaceful, they had to give them tidbits and interpret subtle behaviors such as turning its head and looking toward or stepping toward another bird as "dominant".[34] In another part of their book, they describe naturally occurring aggressive behavior—birds locking in combat and pecking with forceful stabs—but these behaviors were not of interest to their quest for the alpha male, because they were performed by females. Instead, they described female aggression as "the bird version of PMS",[35] which they call PBS (prebreeding syndrome) and did not give it any further attention. Hence, Lawton and colleagues' gender analysis shows that female aggression was dismissed as unimportant and no questions were asked about the possible benefits for females or effect on dominance hierarchies. But they also suggest another interpretation based on Marzluff and Balda's observations which explains them better, namely that female pinyon jays compete for breeding territories and that female rank determines that of their mates.

Here, the theoretical focus on males made the researchers oblivious to the actions and behavior of females. Likewise, in studies of animal behavior from ecological and evolutionary perspectives at large, the vast majority of both theoretical and empirical research has focused on male behavior, leaving females mostly unstudied.[36] This pattern is related to the currently prevailing theoretical framework in sexual selection, also called the Darwin-Bateman paradigm,[37] which predicts males to be competing among each other in relation to reproduction to a much greater

[33] Marzluff & Balda, 1992.

[34] Lawton et al., 1997:70.

[35] Ibid: 71. PMS is the denomination of the premenstrual syndrome, including mood swings and other symptoms.

[36] Fausto-Sterling et al., 1997.

[37] Dewsbury, 2005.

extent than females. This assumption of lesser or non-existent sexual selection in females is one reason why the research focus in the field has been biased toward males.[38]

Feminist biologists and science scholars have criticized these androcentric assumptions, proposed new hypotheses and conducted studies with gender-reflective approaches. This criticism has, to a greater or lesser extent, led to the inclusion of research on females and gender perspectives in different scientific fields.[39] Evolutionary biologists today generally acknowledge the impact of Victorian ideals on Darwin's descriptions of females and women.[40] Yet, some researchers claim that gender bias belongs to the history of sexual selection,[41] at the same time as new feminist critiques continue to emerge.

## Theoretical Framework

Science philosopher Helen Longino argues that scientific knowledge is inherently social, because it is constructed by "an interactive dialogic community", that is "by individuals in interaction with one another in ways that modify their observations, theories and hypotheses, and patterns of reasoning".[42] Thus, a researcher might make a discovery, but that discovery becomes knowledge only upon critical reception in the scientific community. Some findings are immediately accepted by the community, others are met with resistance and give rise to controversies. Because of these controversies, a scientific community's consensus is not a mere reflection of the progress of new theory or objectively discovered empirical findings, but a result of intensive negotiation of scientific knowledge.[43] Also, as feminist biologist Ruth Hubbard states: "though nature is material and real, our descriptions and understandings of it are

[38] Hrdy, 1986, Gowaty, 1997, Dewsbury, 2005, Tang-Martínez & Ryder, 2005.

[39] Haraway, 1989, Fedigan, 2001, Nelson, 2017.

[40] Parker & Birkhead, 2013:2, Zuk & Simmons, 2018:7.

[41] For example Parker & Birkhead, 2013.

[42] Longino, 2006: 272.

[43] Kuhn, 1962, Keller, 1992, Gieryn, 1999.

necessarily mediated by the culture of science".[44] Culture can come to structure scientific theories and practices, however usually not, as Longino has ascertained, as a willful imposition against the evidence, but through the way questions are posed and data interpreted.[45] This book is a study of science as a cultural and social process. Drawing on feminist science studies, I explore how new theories and findings about females were developed and received in the scientific community of sexual selection. I use *Epistemology of Ignorance*[46] (also called *Agnotology*[47])—the study of how and why diverse forms of knowledge are not formed, ignored or delayed—to explore the knowledge/ignorance production about females in sexual selection. Furthermore, I employ science theorist Donna Haraway's concept *situated knowledges* to analyze how the researchers' particular knowledges, perspectives and experiences have formed their knowledges/ignorance about females. In my analysis, I put epistemology of ignorance and situated knowledges in close connection, situating the researchers' knowledges—and ignorance—to understand why they knew or did not know about active females. Thus, the specific researchers' situated knowledges explain how ignorance has been produced or resisted, and led to the multifaceted shifts in perceptions about females that constitute the female turn.

Using the turn metaphor to describe historical processes has been criticized for making tendencies seem more homogeneous than they actually were.[48] Similarly, feminist theorist Clare Hemmings has criticized the use of the wave metaphor in the writing of the history of feminist theory, because it emphasizes a progress narrative of change which makes invisible historical multiplicities.[49] Since perceptions about females shifted in many different subfields, I consider it motivated to call this (still ongoing) process a (multifaceted) female turn. However, by writing the detailed history of ignorance production and the researchers' various

---

[44] Hubbard, 1996: 169.

[45] Longino, 1990.

[46] Mills, 1997, Tuana, 2004, Fehr, 2007, Proctor & Schiebinger, 2008.

[47] Proctor & Schiebinger, 2008.

[48] Wilder, 2012.

[49] Hemmings, 2011.

situated knowledges, I highlight multiplicity and variability, thus continuity as well as change in the history of sexual selection.

Boundaries drawn around gender and science have both excluded and marginalized women for a long time, and fostered cultural perceptions of science as masculine.[50] Thus, both women and what has been considered feminine have been omitted from the realm of science—producing ignorance about females. Feminist historians of science, science theorists and scientists themselves have challenged these omissions, and this book joins their efforts.

## Epistemology of Ignorance

*Epistemology of ignorance* focuses on how and why diverse forms of knowledge are ignored, delayed or not formed. In contrast to epistemology, the branch of philosophy concerning knowledge or the study of "how we know", philosopher Charles Mills introduced *epistemology of ignorance* to describe the construction and persistence of ignorance about race. In *the Racial Contract* (1997), Mills depicts the ways in which ignorance about issues of race has been produced and perpetuated, and how epistemology of ignorance supports white privilege through allowing whites to *not* know about racism, racial domination and exploitation of nonwhites. In a similar line, literary theorist Eve Sedgwick, in *Epistemology of the Closet*, examines the relationship between oppression and ignorance regarding homo/heterosexuality.[51] Likewise, science historian Londa Schiebinger emphasizes that "Ignorance is often not merely the absence of knowledge but an outcome of cultural and political struggle".[52] In her study of colonial science, Schiebinger shows that only certain knowledge that was gained in the colonies was transferred back to Europe. The peacock flower (*Poinciana pulcherrima* or *Caesalpinia pulcherrima*) was imported, but the knowledge about the plants' effectiveness as abortifacients (a medicine that induces abortion) was not transferred along with the plants to

[50] Gieryn, 2011: 23ff.

[51] Sedgwick, 1990.

[52] Schiebinger, 2004:237.

Europe—since knowledge about its use as contraception was suppressed by the mostly male colonial scientists as the colonial states were interested in increasing European population growth.[53] Thus, knowledge about abortifacients was lost. Schiebinger concludes: "Nature, after all, is infinitely rich and variable. What we know or do not know at any one time or place is shaped by particular histories, local and global priorities, institutional and disciplinary hierarchies, personal and professional myopia and much else as well."[54]

Thus, ignorance can be actively produced. Science historian Robert Proctor showed how uncertainty about the detrimental health effects of tobacco has been actively maintained by the tobacco industry—through research that contravene scientific consensus, questioning epidemiology, claiming that animal experiments did not reflect human conditions and constantly calling for more research—thereby keeping "the tobacco controversy" alive.[55] In a similar manner, climate change deniers have produced ignorance about the effects of human activities on global warming.[56]

In line with epistemology of ignorance, what evolutionary biologists know and do not know about females at a particular time in history is shaped by the history of the field, research priorities, available technologies, theoretical assumptions, who is included/excluded from the production of science and the situated knowledges of the researchers involved. As feminist philosopher Nancy Tuana showed in her investigation of the history of knowledge/ignorance about the clitoris and the female orgasm, values and politics are intimately connected to what we attend to and what we ignore. When women's orgasms were thought to be essential to reproduction, seventeenth-century Italian anatomist Renaldus Columbus investigated the clitoris and its connection to women's orgasm. Likewise, the women's health movement in the 1970s and forward explored the form and function of the clitoris because they wanted to liberate women's knowledge about their bodies from sexism in the medical literature and to increase women's pleasure. But in between those occasions, women's

[53] Schiebinger, 2008: 155ff.

[54] Ibid: 237.

[55] Proctor, 2008: 11f.

[56] Oreskes & Conway, 2008: 55-89.

pleasure was thought to be unimportant for reproduction, and therefore the clitoris, if it was at all depicted in anatomical portrayals, was merely a small external uninteresting knot.[57] Hence, there is a connection between the interest, context and experiences of researchers and their knowledge, as well as to their ignorance. Ignorance too is situated.[58]

This epistemology of knowledge/ignorance of females in sexual selection outlines who knows and who does not know and why. How is ignorance sustained, cultivated or allowed?[59] Both male and female animals have always been present in the empirical material, but the prevailing theoretical framework focused on how certain males get more mates and offspring than others, through male-male competition or female choice, which emphasized sexual selection in males and led to the assumption that sexual selection in females is less important.[60] Feminist biologists, just as the Women's Health Movement in Tuana's analysis, exert epistemological resistance through undermining the production of ignorance about females and creating liberatory knowledges. Sexual selection theory concerns humans as well as animals, feminist biologists too "identify, critique and transform ignorance".[61] However, as we will see, not only self-described feminists have brought light on females in this field. Science historian and philosopher Sarah Richardson points out the difference between identifying as a feminist (acknowledging that there are gender inequalities in society and wanting to change that) and engaging in gender-reflective practices of making visible how gender conceptions operate in knowledge and incorporate these insights into the practice of scientific research.[62] Thus, irrespective of whether an individual identifies as a feminist or not, they may use gender-reflective practices. Furthermore, my analysis shows that there are a number of different routes by which researchers come to see females as active and multiply

---

[57] Tuana, 2004: 200ff.

[58] Tuana, 2006:3.

[59] Tuana, 2004:196.

[60] Gowaty, 1997: 352-53.

[61] Tuana, 2006:1.

[62] Richardson, 2013: 17. Not to confuse with the common self-description of trans-exclusionary radical feminists as "gender critical".

mating—theoretical, methodological, through studies of specific animals as well as gender-reflective practices in science.

The boundaries of sexual selection, what is included and excluded at any specific time, is an effect of social negotiation of scientific knowledge. As we shall see, these boundaries have gradually expanded to include more active, multiple mating and variable females.

## Situated Knowledges

Natural science methodology aims at scientific objectivity, that is for the results of science not to be influenced by for example values or personal interests. Yet, the traditional view of scientific objectivity as value-neutral and unbiased has been criticized by (among others) feminist science theorists and historians. Certain groups of people, based on gender, class, sexuality and race, have for a long time been marginalized or excluded from participating in the making of science, and these exclusions influence which questions have been asked, methods used and interpretations made.[63] Thereby, prevailing assumptions about sex and race have been shielded from critical scrutiny.[64] Science theorist Donna Haraway ascertains that it is not possible to conduct research through a "god trick of seeing everything from nowhere",[65] because it is not possible to detach researchers from their social and cultural context. Instead, she suggests an alternative feminist objectivity by acknowledging that all knowledges are situated—that is partial, historically context-bound and based in lived experience.[66] In other words, the researchers' "specific *ways* of seeing"[67] and "prosthetic technologies interfaced with our biological eyes and brains"[68] form their particular situated knowledges. This theory of knowledge (epistemology) is my theoretical point of departure. In addition, I use *situated knowledges* as an analytical lens to understand the different

[63] Harding, 1986, Haraway, 1988, Schiebinger, 1989.

[64] Longino, 1990: 78-79.

[65] Haraway, 1988: 581.

[66] See also Sandra Harding 1993 on "Strong objectivity".

[67] Haraway, 1988: 583.

[68] Ibid: 589.

researchers' partial knowledges, and as an approach to make visible my own knowledge-making as situated (See below: *Relations to Interviewees and Impact on Knowledge-Making* and Chap. 7). My analysis shows how the researchers' situated knowledges about sexual selection are formed by their theoretical perspectives, how human senses and physiology affect studies of animals, their study species, their sex/gender[69] and lived experiences. In my interviews, the researchers often contextualize their own and others' knowledge to explain differences in perspectives, mainly in terms of which animals they are studying and which theoretical perspectives they are using. Hence, they talk about their own and other's knowledges as situated, although they do not use that term.

In the Social Sciences and Humanities, situated knowledges is widely acknowledged and is often alluded to in terms of the researcher's social/cultural/political situatedness. An often under-appreciated aspect of the concept of situated knowledges is the active role of the research object. Haraway writes "Situated knowledges require that the object of knowledge be pictured as an actor and agent, not as a screen or a ground or a resource".[70] In my analysis, I concretize how that agency takes form, that the study organism through (in interaction with) the researcher's particular observation/experiment/methodology forms the researcher's specific situated knowledges.

Throughout this book, I show that comprehending the situated knowledges of the different researchers is significant for understanding the process of shifting perceptions about females. By situating the researchers' knowledges and ignorance, I explain who knew about active females and who did not. Different situated knowledges lead to ignorance or insight about female sexual agency and the analysis shows how new combinations of people/politics/science/organisms enabled this intellectual change in evolutionary biology.

---

[69] Within biology, sex and gender are sometimes used interchangeably and sometimes given separate meanings. Within gender/women's studies, the term gender was formulated to differentiate between biological sex differences and socially constructed gender. However, many gender studies scholars have questioned the dichotomy between sex and gender, our knowledge about sex is informed by constructions of gender and socially constructed gender influences biological sex differences, such as muscle mass. Since the distinction is not clear from either field of study, I here use the term sex/gender.

[70] Haraway, 1988: 592.

## Materials

This book is based on fifteen in-depth interviews with researchers among the worldwide community of evolutionary biologists, correspondence with seven additional researchers and the scientific record of sexual selection. The interviews concerned their subjective perspectives of historical events and individual experiences, in other words, these are oral history interviews.[71] In order to get a diversity of perspectives on shifting perceptions about females I chose to interview pioneers presenting new theories, hypotheses or empirical evidence, leading researchers in sexual selection, and researchers working with explicit gender-reflective approaches, including experts on sexual selection in primates, birds, insects, snakes, frogs, lizards, fishes and spiders (see Appendix 1). I employ my theoretical framework to analyze these different accounts and collate the history of shifting perceptions about females and how ignorance about females has been produced and challenged in sexual selection research.

## Relations to Interviewees and Impact on Knowledge-Making

In relation to my interviewees, I am a younger female colleague in evolutionary biology with a PhD in Zoology[72] from Stockholm University, Sweden, but also someone with a reputation as a feminist biologist, proponent of gender perspectives in biology and a current academic affiliation with gender studies. During my postdoctoral research at the Centre for Gender Research at Uppsala University, I investigated sex from an evolutionary perspective and explored gender perspectives in evolutionary biology. Although sex is often considered to be binary and static, there are many examples of organisms that change sex, or have environmental sex determination. A crocodile's egg is unsexed when laid, but

[71] Portelli, 2006:36-37.

[72] My thesis (Ah-King, 2003) "Phylogenetic analyses of parental care systems" traced the evolution of how animals take care of their young in fish, birds and Callitrichine monkeys.

develops into a sex depending on the ambient temperature during egg development. Biologists define sex by the production of eggs or sperm, but the social environment of a Molly fish determines whether it changes sex and starts producing sperm instead of eggs. From my evolutionary perspective, sex varies over evolutionary time and in relation to environmental and social circumstances—sex and sexual traits are dynamic.[73] In the same line of interdisciplinary research, in gender studies and evolutionary biology, I examined how the term "sex roles" is used in animal behavior,[74] analyzed gender bias in genital evolution research[75] and reviewed the flexibility of female and male mate choice.[76] This interest in sex, gender and evolutionary biology is also my incentive for investigating how researchers have theorized and investigated females in sexual selection, my searchlight in the present study.

Some of the researchers I interviewed I had already met, for example at conferences, others I had collaborated with, for example Patricia Gowaty had been the host for my postdoc in Los Angeles, seven years prior to starting the project. As I sent out my requests for interviews, the research platforms academia.edu and researchgate.net reported the google searches and article reads following the tracks of my invitations. Sometimes my interviewees commented on my work in the interviews, hence they were also informed about my previous work. Thus, I am both an insider, an evolutionary biologist, and an outsider, somebody studying the field, and sliding in and out between these positions, both looking "in from the outside and looking out from the inside".[77] I have experienced being bonded with as a fellow evolutionary biologist, as someone also interested in the "female phenomenon", and a fellow feminist biologist, almost belonging to a genealogy of feminist biologists. But I have also been challenged and questioned, and at one time an interviewee anticipated us having different perspectives and wrote to me prior to the interview: "I suspect we may possibly have different views on some issues, but science

[73] Ah-King & Nylin, 2010.

[74] Ah-King & Ahnesjö, 2013.

[75] Ah-King et al., 2014.

[76] Ah-King & Gowaty, 2016.

[77] Trinh, 1998: 418.

is like that, and there is no reason not to discuss things openly and with goodwill!"

Hence, I am doing what anthropologists call "studying up"[78], turning my research gaze toward those who are seniors in my field, with (probably) more influence in the field than myself. Furthermore, I am an active participant in the field since I have reviewed papers on sexual selection during the period of this project and through the pursuit of the project.

In relation to my interviewees, my position could have influenced whom I got access to interview and what my interviewees chose to tell me. Most interviewees immediately agreed to my request for an interview, but I have also been declined. Sometimes in the interviews, ideas arose about what I as a feminist researcher was assumed to be looking for. One such idea mentioned by an interviewee was "male-biased testosterone-poisoned male scientists". This "third presence" sometimes incited challenges to my questions and concerns that I should make a scientifically accurate investigation.

In relation to my knowledge-making, my position thus affects my collection of materials, but also my interpretation and analysis. Moreover, it is important to reflect on which silences are present in my material and how my research position and questions give rise to such silences.[79] It is possible that my interviewees, who certainly want to be perceived as good scientists, are more willing to talk about ways in which they have promoted new perceptions of females, rather than instances when they have held on to old perceptions or resisted new ideas that are currently widely accepted.

The interviews provide different accounts of researchers experiences and knowledge-making, and different perspectives of historical events. I use these accounts to analyze the history of production of knowledge/ignorance about females in sexual selection. Thus, I choose what to include and exclude in writing the history of the female turn. Writing

[78] Nader, 1972.

[79] Farahani, 2010.

history of a field always reduces heterogeneity of reality,[80] and is in itself a negotiation of knowledge.[81]

## Outline of the Chapters

To illustrate the female turn in the different evolutionary subfields, three chapters are categorized according to subfields—Chaps. 2 (primates), 4 (birds) and 5 (insects, snakes, frogs, lizards, fishes and spiders). Furthermore, shifting perceptions about female agency was crucial in the turning appreciation of female choice as a mechanism of sexual selection (Chap. 3), and in the scientific community's development and reception of male- and female-focused hypotheses about sexual selection after mating (Chap. 6). Thus, the chapters are thematic and not presented in a chronological order.

Chapter 2, "An Early Female Turn in Primate Research", concerns the shifting views on females in the research on monkeys and apes. Already in the 1950s, Japanese primatologists recognized females as powerful in Japanese macaque society, and western primatologists' delay in recognizing these findings is one example of ignorance production in sexual selection. In contrast early western[82] primatology was male-focused, primarily investigating male aggression and dominance patterns. Females were mainly considered in their role in procreation and viewed as passive resources for males. The findings that female primates often initiate mating with several males went against theoretical assumptions about "coy" and monogamous females. Feminist primatologists were actively challenging stereotypic perceptions of females, leading to a broader impact on primatology. Compared to the other subfields in evolutionary biology, primatologists at large were early in their investigations of females' lives and strategies.

[80] Nyström, 2015:34; Hemmings, 2011.

[81] Haraway, 1989: 126.

[82] Here, I follow Haraway's practice of not capitalizing "the west" to disrupt its ideological position as "West is One". Primatology has been international from the start, it emerged at about the same time in Japan, Europe and the US, and researchers from other parts of the world have increasingly become involved in primate research (Haraway, 1989: 116).

Ever since Charles Darwin's time, female choice as a mechanism of sexual selection in animals has been the target of controversy, and Chap. 3, "Turning Appreciation of Female Choice", concerns the history of this controversy from its beginning until its end around 1980. Because Darwin's contemporary biologists did not accept the suggestion of female choice in animals, due to their belief that animals did not have the aesthetic sense to exercise choice as well as their Victorian perceptions of sexually passive females—sexual selection in animals was not acknowledged in the nineteenth century. However, mate choice in humans was readily accepted among eugenicists and others as a way to improve humanity. After the turn of the century, there was, with few exceptions, sparing interest in sexual selection in animals until its forceful revival in the 1970s. Feminists contemporary to Darwin pointed out the ignorance of females in his theory and writings about sex differences in animals and humans, but their critique and they themselves became excluded from science with the professionalization and masculinization of the scientific establishment—which contributed to ignorance production about females. Later, the social and political context of the "second wave" feminist movement and the entrance of many women into science set the stage for the vigorous revival of sexual selection. Furthermore, around 1980, new mathematical models showed that female choice could lead to evolution of male elaborate characteristics, and empirical evidence emerged showing female choice in action. Yet, males were the primary focus of sexual selection research at this time, and female choice studies were largely conducted to understand the evolution of male traits. The long controversy over whether female choice could occur as a mechanism of sexual selection ended, however, stereotypical notions of females as passive and coy lingered.

Chapter 4 concerns "A Female Turn in Bird Research". Females were for a long time assumed not to gain any benefits from mating with several males, but the discovery of frequent female multiple mating led to what some researchers call the "polyandry revolution",[83] that is perceptions of females shifted from monogamous to often mating with multiple males. Birds are often described as the classical animal group of sexual selection

[83] Parker & Birkhead, 2013.

research, and bird research was the primary scene for the “polyandry revolution”. New molecular methods for determining parentage revealed surprisingly widespread female multiple mating among widely presumed monogamous birds. These findings were initially interpreted as due to males imposing copulations on unwilling or passive females, but later studies showed that females can actively initiate multiple mating. Certain researchers, due to their specific situated knowledges—as feminists, due to their engagement in investigating females or to their specific knowledge of particular species—were crucial for putting forward interpretations of females as active, variable, aggressive, dominant and worth studying in their own right.

In this epistemology of knowledge/ignorance about females in sexual selection, Chapter 5—"Shifting Perceptions About Female Insects, Frogs, Snakes, Lizards, Fishes and Spiders”—shows that there are many different routes toward knowledge about active females and female multiple mating. Among insect researchers, females were known to mate with multiple males long before DNA fingerprinting technology—which was crucial in studies on birds and other animals—this knowledge was transferred from pest-control studies, natural history observations and population genetics. An example from frog research shows how interdisciplinary methodology—combining neurobiology, animal behavior and sexual selection—engendered an emphasis on female agency. In fish research, feminist researchers highlighted the role of females in sexual selection. Furthermore, research on sexual selection in spiders has its own idiosyncratic history, focusing primarily on the large, sometimes sexually cannibalistic, females. Lastly, I discuss taxonomic ignorance and asymmetric knowledge circulation in sexual selection.

In Chap. 6, “Sexual Selection After Mating: Turning Perceptions of Female Agency”, I investigate the scientific community’s development and reception of the equivalent male and female hypotheses about sexual selection after copulation: sperm competition (competition among different males’ sperm for fertilization of eggs) and cryptic female choice (females influencing which sperm fertilizes their eggs during or after copulation). Initially, there was a polarized controversy between researchers with these different perspectives. Given that sperm competition expanded sexual selection to a new temporal and spatial site—after copulation and within the female reproductive tract—one could have imagined both male-focused and female-focused ideas simultaneously, but sperm competition preceded cryptic female choice in the minds of evolutionary biologists. Still, sperm competition is much more explored than cryptic female choice, which is often more difficult to study in practice and frequently remains overlooked.

In Chap. 7, "Conclusions: An Epistemology of Ignorance in Sexual Selection", I summarize the recent history of sexual selection from a gender perspective. Perceptions about females have gradually shifted and continues to change from coy, passive and monogamous toward acknowledging that females can have active sexual strategies and often mate with multiple males—an ongoing *female turn* in perceptions about females. These perceptions changed at different times in the sub-fields and epistemic communities focusing on different animals depending on the specific circumstances—prior knowledge about females from other fields of research, particular assumptions associated with the animals in question, development of new technologies and the partial context-specific knowledge and lived experiences of individual researchers—including feminist insights. The analysis demonstrates a repeated pattern of male precedence in sexual selection studies, a kind of androcentrism or ignorance about females—when research starts with male-centered investigations or explanations and later includes the female-centered equivalents.[84] The ignorance production about females in sexual selection is both historical and ongoing. Analyzing the history of shifting perceptions about females (and males) also sheds light on what continues to be neglected in sexual selection research.

## References

Ah-King, M. (2003). *Phylogenetic analyses of parental care evolution*. Department of Zoology, Stockholm University, Sweden.

Ah-King, M. (2022). The history of sexual selection research provides insights as to why females are still understudied. *Nature Communications 13*(1) 6976 http://doi.org.10.1038/s41467-022-34770-z

[84] Ah-King 2022.

Ah-King, M., & Ahnesjö, I. (2013). The "sex-role" concept: A review and evaluation. *Evolutionary Biology, 40*, 461–470.
Ah-King, M., & Gowaty, P. A. (2016). A conceptual review of mate choice: Stochastic demography, within-sex phenotypic plasticity, and individual flexibility. *Ecology and Evolution, 6*(14), 4607–4642.
Ah-King, M., & Nylin, S. (2010). Sex in an evolutionary perspective: Just another reaction norm. *Evolutionary Biology, 37*, 234–246.
Ah-King, M., Barron, A., & Herberstein, M. E. (2014). Genital evolution: Why are females still understudied? *PLOS Biology, 12*(5), e1001851.
Andersson, M. (1994). *Sexual selection*. Princeton University Press.
Bateman, A. J. (1948). Intra-sexual selection in Drosophila. *Heredity, 2*, 349–368.
Blackwell, Antoinette Brown. (1875). Republished in 1976. *The sexes throughout nature*. Putnam.
Bleier, R. (1984). *Science and gender: A critique of biology and its theories on women*. Pergamon Press.
Burke, Meghan A., & Embrick, David G. (2020). Biological determinism. In: *International Encyclopedia of the Social Sciences*. https://www.encyclopedia.com/science-and-technology/biology-and-genetics/biology-general/biological-determinism
Darwin, Charles. (1860). Letter to Asa Gray 3 April 1860. *Darwin Correspondence Project*, "Letter no. 2743,". Retrieved April 3, 2020, from https://www.darwinproject.ac.uk/letter/DCP-LETT-2743.xml
Darwin, Charles. (1871). Republished in 1981. *The descent of man and selection in relation to sex*. Reprint of the 1871 edition published by J. Murray, London. Princeton University Press.
Darwin, Charles. (1859) Republished in 2006. *On the origin of species*. Dover Publications.
Dewsbury, D. (2005). The Darwin-Bateman Paradigm in historical context. *Integrative and Comparative Biology, 45*, 831–837.
Evans, Samantha (compiler). (2017). *Darwin and women: A selection of letters*. Cambridge University Press.
Farahani, F. (2010). On being an insider and/or an outsider: A diasporic researcher's catch-22. In L. Naido (Ed.), *Education without borders: Diversity in a cosmopolitan society* (pp. 113–130). Nova Science Publisher Inc.
Fausto-Sterling, A., Gowaty, P., & Zuk, M. (1997). The moral animal: Why we are the way we are: The new science of evolutionary psychology—Wright, R. *Feminist Studies, 23*, 403–417.
Fedigan, L. M. (2001). The paradox of feminist primatology: The Godess's discipline? In A. N. H. Creager, E. Lunbeck, & L. Schiebinger (Eds.), *Feminism*

*in twentieth-century science, technology, and medicine*. University of Chicago Press.
Fehr, C. (2007). Are smart men smarter than smart women? The epistemology of ignorance, women and the production of knowledge. In A. M. May (Ed.), *The 'Woman Question' and higher education: Perspectives on gender and knowledge production in America*. Edward Elgar Publishing Inc.
Gieryn, T. F. (1999). *Cultural boundaries of science: Credibility on the line*. University of Chicago Press.
Gieryn, T. F. (2011). Boundaries of science. In S. Jasanoff, G. E. Markle, J. C. Peterson, & T. Pinch (Eds.), *Handbook of science and technology studies*. SAGE Publications, Inc..
Gowaty, P. A. (1997). Sexual dialectics, sexual selection and variation in reproductive behavior. In P. A. Gowaty (Ed.), *Feminism and evolutionary biology, boundaries, intersections and frontiers* (pp. 351–384). Chapman and Hall.
Gowaty, P. A., Kim, Y.-K., & Anderson, W. W. (2012). No evidence of sexual selection in a repetition of Bateman's classic study of *Drosophila melanogaster*. *Proceedings of the National Academy of Science USA, 109*(29), 11740–11745.
Grosz, E. (2011). *Becoming undone: Darwinian reflections on life, politics, and art*. Duke University Press.
Hamlin, K. A. (2014). *From eve to evolution: Darwin, science, and women's rights in Gilded Age America*. The University of Chicago Press.
Haraway, D. J. (1988). Situated knowledges: The science question in feminism and the privilege of partial perspectives. *Feminist Studies, 14*(3), 575–599.
Haraway, D. J. (1989). *Primate visions: Gender, race and nature in the world of modern science*. Routledge.
Harding, S. (1986). *The science question in feminism: Industrial policy in Europe*. Cornell University Press.
Harding, S. (1993). Republished in 2006. Rethinking standpoint epistemology: What is 'strong objectivity'? In: Keller, Evelyn F. & Longino, Helen E. (eds.) Feminism & Science, pp. 235-248. Oxford University Press.
Hemmings, C. (2011). *Why stories matter: The political grammar of feminist theory*. Duke University Press.
Hrdy, S. B. (1981). *The woman that never evolved*. Harvard University Press.
Hrdy, S. B. (1986). Empathy, polyandry and the myth of the coy female. In R. Bleier (Ed.), *Feminist approaches to science*. Pergamon Press.
Hrdy, S. B. (2000). *Mother Nature: Maternal instincts and how they shape the human species*. Ballantine Books.
Hubbard, R. (1996). Gender and genitals: Constructs of sex and gender. In A. Ross (Ed.), *Science wars*. Duke University Press.

Keller, E. F. (1992). *Secrets of life, secrets of death: Essays on language, gender and science*. Routledge.
Kuhn, T. S. (1962). *The structure of scientific revolutions*. University of Chicago Press.
Lawton, M. F., Garstka, W. R., & Hanks, J. C. (1997). The mask of theory and the face of nature. In P. A. Gowaty (Ed.), *Feminism and evolutionary biology, boundaries, intersections and frontiers* (pp. 63–85). Chapman & Hall.
Longino, H. E. (1990). *Science as social knowledge: Values and objectivity in scientific inquiry*. Princeton University Press.
Longino, Helen E. (2006). Subjects, power, and knowledge: Description and prescription in feminist philosophies of science. In: Keller, Evelyn F. And Longino, Helen E. (eds.), *Feminism and science*, pp 264-279. Oxford University Press.
Marzluff, J., & Balda, R. P. (1992). *The Pinyon Jay: Behavioral ecology of a colonial and cooperative corvid*. Poyser.
Milam, E. L. (2010). *Looking for a few good males: Female choice in evolutionary biology*. John Hopkins University Press.
Mills, C. W. (1997). *The racial contract*. Cornell University Press.
Nader, L. (1972). Up the anthropologist: Perspectives gained from studying up. In D. Hymes (Ed.), *Reinventing Anthropology*. Pantheon Books.
Nelson, L. H. (2017). *Biology and feminism: A philosophical introduction*. Cambridge University Press.
Nyström, D. (2015). *Innan forskningen blev radikal: en historiografisk studie av arbetarhistoria och kvinnohistoria*. Universus Academic Press.
Oreskes, N., & Conway, E. M. (2008). Challenging knowledge: How climate science became a victim of the cold war. In R. N. Proctor & L. Schiebinger (Eds.), *Agnotology: The making and unmaking of ignorance* (pp. 55–89). Stanford University Press.
Parker, G. A., & Birkhead, T. R. (2013). Polyandry: The history of a revolution. *Philosophical Transactions of the Royal Society B-Biological Sciences, 368*, 1471–2970.
Portelli, A. (2006). What makes oral history different? In R. Perks & A. Thomston (Eds.), *The oral history reader*. Routledge.
Proctor, R. N. (2008). Agnotology: A missing term to describe the cultural production of ignorance (and its study). In R. N. Proctor & L. Schiebinger (Eds.), *Agnotology: The making and unmaking of ignorance* (pp. 1–33). Stanford University Press.
Proctor, R. N., & Schiebinger, L. (Eds.). (2008). *Agnotology: The making and unmaking of ignorance*. Stanford University Press.

Richards, E. (2017). *Darwin and the making of sexual selection.* The University of Chicago Press.
Richardson, S. S. (2013). Sex itself: The search for male and female in the human genome. University of Chicago press.
Saini, A. (2017). *Inferior: How science got women wrong—and the new research that's rewriting the story.* 4th Estate.
Schiebinger, L. (1989). *The mind has no sex? Women in the origins of modern science.* Harvard University Press.
Schiebinger, L. (2004). Feminist history of colonial science. *Hypatia, 19*(1), 233–254.
Schiebinger, L. (2008). West Indian abortifacients and the making of ignorance. In R. N. Proctor & L. Schiebinger (Eds.), *Agnotology: The making and unmaking of ignorance* (pp. 149–162). Stanford University Press.
Sedgwick, E. K. (1990). *Epistemology of the Closet.* University of California Press.
Snyder, B. F., & Gowaty, P. A. (2007). A reappraisal of Bateman's classic study of intrasexual selection. *Evolution, 61-11*, 2457–2468.
Tang-Martínez, Z., & Ryder, B. T. (2005). The problem with paradigms: Bateman's worldview as a case study. *Integrative and Comparative Biology, 45*, 821–830.
Travis, C. (1992). *The mismeasure of woman, why women are not the better sex, the inferior sex, or the opposite sex.* Touchstone.
Trinh, T. M.-h. (1998). Not you/like you: Post-colonial women and the interlocking questions of identity and difference. In A. McClintock, A. Mufti, & E. Shohat (Eds.), *Dangerous liaisons: Gender, nation, and postcolonial perspectives* (pp. 415–419). University of Minnesota Press.
Trivers, R. L. (1972). Parental investment and sexual selection. In B. Campbell (Ed.), *Sexual Selection and the Descent of Man* (pp. 136–179). Aldine.
Tuana, N. (2004). Coming to understand: Orgasm and the epistemology of ignorance. *Hypatia, 19*(1), 194–232.
Tuana, N. (2006). The speculum of ignorance: The Women's Health Movement and epistemologies of ignorance. *Hypatia, 21*(3), 1–19.
Wilder, G. (2012). From optic to topic. The foreclosure effect of historiographic turns. *The American Historical Review, 117*(3), 723–774.
Woolf, Virginia. (1938) Republished in 1966. *Three Guineas.* Harcourt.
Zuk, M., & Simmons, L. W. (2018). *Sexual Selection: A very short introduction.* Oxford University Press.

# 2

# An Early Female Turn in Primate Research

The Japanese macaque, *Macaca fuscata*, also called snow monkey, resides on three of Japan's four main islands, from the evergreen forests in the south to the snow-covered mountains in the north. It is the northernmost-living of all non-human primates and is recognized by its pink face and dense grayish fur. In October 1948, animal ecologists Syunzo Kawamura and Junichirô Itani were studying semi-wild horses at the scenic seaside mountainous Cape Toimisaki when they came across a troop of Japanese macaques, an encounter that impressed and inspired them to expand their studies of animal societies to primates.[1] Japanese researchers aimed at understanding evolution of human society through studying the whole "species society", therefore they kept track of social relations between all members of a group.[2] This holistic perspective on primate society led to their 1958 conclusion that in Japanese macaques, female kinship was an important determinant of dominance hierarchies.[3] Of importance for the development of perceptions about females in primatology, Japanese primatologists discovered female dominance and female agency several years

[1] Takasaki, 2000: 152.

[2] Asquith, 2000: 167.

[3] Kawamura, 1958: 149.



M. Ah-King, *The Female Turn*, https://doi.org/10.1007/978-981-19-7161-7_2

prior to their western colleagues. Since primate studies were developed independently in Japan and in western countries within a decade of each other (in the 1940s versus 1930s)—these differences make visible cultural influences on knowledge production in science.[4] Unlike western primate studies, Japanese primatology doesn't seem to have undergone a female turn, because, in contrast to (most) western[5] primatologists, they did not start out assuming that society revolved around active males and that females were passive. Later in this chapter, I will get back to how the Japanese approach to studying primates led them to consider females as powerful and active early on.

Because primates are our closest living relatives, studies of them have been used to speculate about evolution and behavior in humans, thus, both researchers and the general public have shown exceptional interest in primate studies.[6] Furthermore, views about human society are reflected in the research of monkeys and apes.[7] Primatology is an interdisciplinary field which brings together anthropologists, animal behaviorists, ecologists, zoologists and psychologists studying monkeys and apes. Science theorist and historian Donna Haraway analyzed the development of primatology and demonstrated that western scientific narratives are deeply entrenched in western cultural dichotomies of nature/culture, gender/sex, male/female, human/animal, subject/object.[8] She highlights that primatology is a major area of feminist concern because primate science is important in debates about human—and especially female—nature.[9] Her analysis showed that primatology is a field replete with controversy in which scientists contest the origins of human nature, "the family" and

[4] Asquith, 1996: 241.

[5] Donna Haraway points out that although European and Euro-American researchers have dominated western primatology, it has always been cacophonous and contested. Hence, I follow Haraway's practice of not capitalizing "the west" to disrupt its ideological position as "West is One". History of primatology is international, it emerged at about the same time in Japan and Europe and the US, and researchers from Latin America, Asia and Africa have increasingly become involved in primate research (Haraway, 1989: 116).

[6] Strum & Fedigan, 2000: 5.

[7] Haraway, 1989; Asquith, 1996: 241.

[8] Haraway, 1989: 82.

[9] Ibid.: 77.

society, and in all these origin stories, woman/female is the marked category, the "other" to the male norm.[10]

Many science analysts have noted a connection between the increase in women in primatology, the emergence of feminist critique and changes in the scientific content and methodology of primatology. Therefore, a number of feminist scholars have investigated the development of this field—Donna Haraway, anthropologist Pamela Asquith, science historian Londa Schiebinger, science philosopher Lynn Hankinson Nelson, primatologists Sarah Blaffer Hrdy, Linda Fedigan and Shirley Strum.[11] In this chapter, I draw on this earlier research and my interviews with Sarah Blaffer Hrdy and Meredith Small, as well as the scientific record, to investigate the female turn in primate research.

This chapter shows that western primatology has shifted perceptions about females in multiple ways—from the initial assumptions of being coy, passive, peaceful, disinterested in dominance and not experiencing orgasms to sexually active, aggressive, highly involved in dominance hierarchies and experiencing orgasms. I describe how these shifts came about through feminist critique, changes in methodology, language use, and model species for human evolution. I present the experiences and perspectives of the two interviewed primatologists, and discuss the shifting perceptions about females in the closely related field of anthropology. Thereafter, I consider the early recognition of active females in Japanese primatology, the role of anthropomorphism (attribution of human characteristics to animals) and empathy, how primatology can be seen as politics,[12] and the role of women, feminism and gender-reflective approaches[13] in the field. This chapter investigates how the different researchers' situated knowledges[14] and ignorance have formed their understandings of females and sexual selection. Thereby the analysis shows who did and who did not know about active females and when, and demonstrates several different ways in which ignorance has been pro-

[10] Ibid.: 83.

[11] Haraway, 1989, Asquith, 1996, 2000, Schiebinger, 1999, Nelson, 2017, Hrdy, 1986, Fedigan, 2001, Strum & Fedigan, 2000.

[12] Haraway, 1986, 1989.

[13] Richardson, 2013: 17.

[14] Haraway, 1988.

duced. Here follows the multifaceted history of the female turn in primate studies.

## Early Western Primatology

In western primatology, there has been a dramatic shift in perceptions about females (and males)—from the assumption that primate societies generally revolve around males through aggression and dominance—to a more complex picture of variability in primate social organization in which females play an active role, and both males and females can compete and have dominance hierarchies.[15] There has been a repeated pattern of male precedence—when research starts with male-centered investigations or explanations and later includes the female-centered equivalents, a kind of androcentrism (the practice of placing the male point of view in the center of one's world view).[16] This history of production of ignorance about females starts in the male-dominated scientific community of the early western primatologists. Both Donna Haraway and Sarah Hrdy have pointed out the connection between early male-dominated western primatology and focus on male-male competition, and the subsequent shift in focus to female behavior when many women in the beginning of the 1970s had entered primatology. Feminist science scholarship argues that the widespread androcentrism in biological and social sciences was due to the male dominance and exclusion of women from these fields, and that the situatedness of researchers in terms of their sex/gender influences their knowledge.[17] I will get back to the role of women, feminism and gender-reflective approaches later in this chapter.

Although early descriptions of primate research were heterogeneous, the prevailing perception of primate society was androcentric and the majority of studies focused on male dominance over females, male dominance hierarchies and male aggression.[18] One of the early male and

[15] Strum & Fedigan, 2000: 5.
[16] Ah-King 2022:
[17] Haraway, 1989, Nelson, 2017: 97.
[18] Fedigan, 1982/1992.

aggression-focused studies was primatologist Solly Zuckerman's influential study *The Social Life of Monkeys and Apes* (1932)[19] on a captive group of hamadryas baboons (*Papio hamadryas*) at the London Zoo. This group of baboons initially consisted of one hundred males in a small area, later joined by some females—in light of today's knowledge a much too large number of individuals, unfamiliar to each other, in a too small space and with an inappropriate sex ratio—the resulting chaos was interpreted as "normal" behavior and Zuckerman concluded that fierce sexual competition between males was the general basis for primate society.[20]

In contrast, contemporary with Zuckerman, American primatologist Ray Carpenter insisted that primate behavior should be studied in their natural habitat and he pioneered primate field studies with a study of howler monkeys (*Alouatta palliata*) in South America in 1934. He found that howler females were more assertive and sexually motivated for mating than males.[21] He described the howler society as built on mutualistic social relationships, with communal sexual relations forming the basis for group cohesion. Male competition and ritual aggression occur between groups, but not within groups and there were no obvious dominance hierarchies among either males or females.[22] In line with the prevailing androcentrism which contributed to ignorance about females, Zuckerman's baboon model became the dominating model for interpreting primate society for decades,[23] and Carpenter even reinterpreted howler society in his later work—in line with the competitive and male-dominated model of baboons.[24]

When American primate field studies were revived after World War II, they were fueled by the quest for understanding human aggression and social stability, largely influenced by Sherwood Washburn at the Anthropology department of the University of California, Berkeley.[25] He and his students believed humans and non-humans shared certain

---

19 Zuckerman, 1932/2015.

20 Strum & Fedigan: 8f.

21 Ibid.: 9.

22 Ibid.: 9.

23 Ibid.: 42.

24 Ibid.: 15.

25 Haraway, 1989: 126.

adaptive features—a "primate pattern", representing primate "nature"—which was assumed to differ little between species and contexts.[26] In line with this theoretical framework, they motivated evolutionary speculations, generalizations and extrapolations from one species to another.[27] Among the 200 species of primates, the primary species for representing a model for human evolution for a long time was the common (or yellow) baboon, *Papio cynocephalus*. The motivation for using them as a model was that they were early to be extensively studied in the field, they are ground-dwelling and adapted to a terrestrial life on the open savannah, just as humans.[28] They are also characterized by male bonding, male dominance hierarchies and high levels of aggression—which was then generalized to represent the "primate pattern" as well as early human evolution. Studies aimed at describing the characteristics of *the* primate male, *the* primate female and *the* immature primate.[29] In contrast to the numerous baboon studies, field studies of langurs in Northern India provided a portrayal of a peaceful primate society—in which females shared caring of infants and males, although having a dominance hierarchy, were peripheral to the group.[30]

Despite the variability in primate social relations among species revealed by these early studies, the baboon model became representative of a general primate pattern and of human evolution during 1950–1965—with male dominance hierarchy, males as leaders and defenders and females caring for infants.[31] Primatologist Linda Fedigan suggests that this model gained popularity because it conformed to the researchers' cultural worldview in the male-dominated field of primatology, because the early descriptions of baboons as male-oriented and military-structured have later been questioned[32] (see below).

Accordingly, it was common in primatology in the 1960s–1970s to describe males as "owning" their females and assuming that females were

[26] Strum & Fedigan, 2000: 10.
[27] Ibid.: 10.
[28] Fedigan, 1982/1992: 308.
[29] Strum & Fedigan, 2000: 10.
[30] Ibid.: 12.
[31] Ibid.: 14.
[32] Fedigan, 1982/1992: 309.

passive resources that males competed for.[33] While today, males in many primate species are understood as attaching themselves to permanent kin groups of females that occupy a territory.[34] Primatologist Sarah Hrdy recalls that at the time that she started studying primates in 1970,[35] the go-to reference on female primates, typically entitled "The Female Primate" by biological anthropologist Phyllis Jay (1963), described females as focused on producing as many babies as they can:

> Her primary focus, a role which occupies more than 70 percent of her life, is motherhood. […] A female raises one infant after another for her entire adult life […] Dominance interaction is usually minimal in the life of a female.[36]

The prevailing idea was also that females did not vary among themselves in reproductive success. A normal female was assumed to be a mother, and a female without offspring was thought to be socially problematic. Furthermore, comparative psychologists in their 1960s experimental laboratory set-ups isolated females with their offspring and thereby precluded observations of any interactions with other group members but their offspring, thereby cementing perceptions of females as merely mothers.[37] Thus, when females were studied in the field, they turned out to have many more facets of behavior. Likewise, Meredith Small points out that in laboratory studies of some South American monkeys, such as marmosets and tamarins (belonging to the family *Callitrichidae*), females were assumed to be monogamous and these assumptions were maintained because nobody ever thought about adding another male into the cage.[38] Thus, females mating with multiple males was discovered much later.

---

[33] Fedigan, 2001: 53.
[34] Ibid.: 53.
[35] Sarah Hrdy interview 2017.
[36] Jay, 1963 cited in Hrdy, 1981/1999: 201.
[37] Hrdy, 2000: 28.
[38] Meredith Small interview 2017.

## Shifting Perceptions About Females

A number of biases—methodological, ideological and theoretical—led western primatologists to believe in assumptions about females as coy and passive along with stereotypical assumptions about males.[39] It was not until the 1970s that females started to become studied in their own right, not as before—merely in their role as mothers, or as passive resources for males.[40] This history shows the production of ignorance about females, but also how this production has been undermined in several different ways—by the critique of assumptions of females as "coy" and passive, peaceful and uninterested in dominance; through shifting methodology, language use, primate models for human evolution and transforming the comprehension of female orgasms. These are different aspects of how perceptions of females have shifted in western primatology.

### Females Are Not Coy and Passive

The discovery that female langurs, macaques, chimpanzees and many other primates are sexually assertive and mate with several different males went against the commonly held assumptions about coy females—especially since primate females invest heavily in pregnancy and lactation they were expected to be particularly choosy in picking a mate. In 1974, Sarah Hrdy reported female multiple mating and variation in female reproductive success in Hanuman langurs (*Semnopithecus entellus*) (see section on Hrdy below). In 1977, Alan Dixon described Celebus monkey females (*Macaca nigra*) actively soliciting matings and mating with multiple males, as well as same-sex sexual behavior among both sexes.[41] When DNA fingerprinting for determining parentage became available, it also revealed that infants of females were not always fathered by the males in the group, even when behavioral observations had given no indications.[42]

[39] Hrdy, 1988: 164.

[40] Haraway, 1989.

[41] Dixon, 1977.

[42] Fedigan, 2001: 58.

Not only researchers of empirical studies, but also authors of theoretical models started to describe females as active. In 1980, Richard Wrangham developed a new model of primate society organization—starting with females distributing themselves in relation to resources, and then males distributing themselves in relation to these female-bonded groups.[43] This model was influential in shifting away from perceptions of females as passive resources to males.[44]

As the listed names indicate, both men and women contributed to shifting these perceptions. Irwin Bernstein, who conducted comparative studies on dominance in different species of captive primates, in 1981[45] pointed out that females cannot be considered as passive resources, likewise did Linda Fedigan in *Primate Paradigms: Sex Roles and Social Bonds* 1982.[46] Donald Lindburg was early (1967) to focus on female sexual choice in rhesus macaques[47] (*Macaca mulatta*) and Meredith Small studied female choice and relationships in different species of macaques from 1970s and onwards (see section on Small below).

Most primate studies had been conducted on Old World monkeys[48] and generalization of aggressive and competitive male primates were based on these, but Karen Strier challenged these generalizations through her studies of New World monkeys[49]—demonstrating that these include affiliative, closely bonded males.[50] Here is one example of how the specific study species is one aspect forming researchers' situated knowledges. Not only have primatologists shown how females play active roles in primate societies, they have also demonstrated that there is a variety of relations between and within the sexes.[51]

43 Wrangham, 1980.

44 Fedigan, 2001: 59f.

45 Bernstein, 1981.

46 Fedigan, 1982/1992.

47 Ibid.: 332, 338.

48 Old world monkeys are the monkey species presently native in Africa and Asia.

49 New World monkeys are the species presently native in Mexico, Central and South America.

50 Fedigan, 2001: 60.

51 Ibid.: 53.

## Female Dominance

Western primate studies started out with assumptions of dominant males and passive females, and early short-term studies did not change this view. It was not until long-term studies were conducted that these assumptions started to shift and researchers realized that in both baboons and macaques, males migrate between groups, but females remain in their natal group and thus female kin relations are at the core of these primate societies.[52] Donald Sade (who himself discounted the influence of the women's movement in primatology[53]) reported matrilinear structure of rhesus macaques from his long-term studies in the 1960s, although he was rarely cited by those emphasizing the prevailing male dominance hierarchy model[54]—an example of production of ignorance through not citing studies with alternative notions. Long-term studies and individual recognition also led Japanese primatologists to recognize female dominance prior to Sade's study[55] (see section on Japanese primatology below).

Assumptions about dominant males became questioned from many different perspectives. In her article "In praise of the achieving female monkey", feminist primatologist Jane Lancaster (1973) pointed out that many primate societies are structured around female matrilines and that much of the knowledge transmission between generations is primarily performed by females.[56] Her study of vervet monkeys (*Chlorocebus pygerythrus*) was the first to report that females are sometimes dominant to males in a primate species. Later, Alison Jolly found that females of several lemur species (e.g. *Lemur catta* and *Indri indri*) can be dominant over males.[57] Here again, it is clear how the study species forms the researcher's situated knowledges and forms the basis to develop alternative models of primate society.

---

[52] Strum & Fedigan, 2000: 23.

[53] Losos cited in Haraway, 1989: 412n22.

[54] Losos cited in Haraway, 1989: 412n22.

[55] Asquith, 1996.

[56] Hrdy, 1981/1999: 201.

[57] Strum & Fedigan, 2000: 25.

Thelma Rowell (1974) pioneered the critique of the male dominance model in baboon society, suggesting that male-dominated hierarchies among scientists as well as conditions induced by captivity influenced this view.[58] Furthermore, she and Vernon Reynolds were early to emphasize that female-female coalitions were important.[59] Shirley Strum's studies of baboons showed that male aggression was not as important for dominance hierarchies as earlier studies suggested, males also have non-aggressive alternative strategies and these social approaches are less risky but as effective in competition and defense.[60] In addition, Shirley Strum and Barbara Smuts demonstrated the importance of friendships between males and females, and between males and infants.[61] Females were also found to be policing, protecting, leading and having stable dominance hierarchies—behavior that was formerly attributed only to males.[62]

In a 1983 review of studies on dominance and reproductive success, Linda Fedigan showed the gender bias in studies on reproductive success.[63] She found 20 studies of male reproductive success and only eight concerning female reproductive success, despite the fact that female reproductive success was much easier to measure accurately at the time. Male reproductive success was approximated by the number of copulations (which is not directly translatable into reproductive success since all matings do not result in fertilization). Furthermore, she emphasized the importance of studying female reproductive strategies—that females too compete among themselves for resources and rank. Thus, shifting understandings of dominance relations contributed to shifting perceptions about females in primatology.

[58] Fedigan, 2001: 59.

[59] Hrdy, 1981/1999: 201.

[60] Strum & Fedigan, 2000: 23.

[61] Ibid.: 24.

[62] Ibid.: 23.

[63] Fedigan, 1983.

## Changing Methodology

One of the strategies that feminist primatologists used to challenge androcentric research was to criticize prevailing methodology. As described above, early western studies focused on male aggression and dominance hierarchies. Feminist primatologist Jeanne Altmann became particularly influential with her 1974 publication "Observational studies of behavior: Sampling methods",[64] in which she evaluated methodology in field studies. She concluded that the common *ad libidum* sampling method—observing animals and noting what catches one's attention—detects rare significant events, but is not useful for investigating questions about rates or durations and does not allow for comparisons between subgroups. She proposed that another way of observation, focal sampling—following and observing what a certain individual does during a predetermined time and shifting focal animals—was a much more useful way of collecting data that could be compared between groups, between sexes and individuals. Such data also foregrounded what the hitherto overlooked females did.[65] Altmann convinced scientists to raise their standards of evidence by methodological improvement,[66] not only in primatology but in animal behavior studies generally. Thus, primatologists moved away from focusing on the large males' aggressive interactions and started paying attention to females, subdominant males and immatures.[67] Altmann's paper became one of the most cited in animal behavior[68] and changed how scientists collected data in animal behavior studies generally. It also destabilized the status of previous primate field research, because it became obvious that many studies published before the mid-1970s had collected and interpreted data in ways that could not be justified.[69] Altmann thus criticized the primatology literature for focusing

[64] Altmann, 1974.

[65] Haraway, 1989: 307.

[66] Fedigan, 2001: 51.

[67] Ibid.: 51.

[68] Haraway, 1989: 305.

[69] Ibid.: 307.

mainly on high drama—sex and violence.[70] Using the focal sampling method enabled reconsiderations of sex issues in primatology—by reducing emphasis on high drama and foregrounding everyday mundane life.

This inclusion of females in study samples happened in parallel to similar shifts in psychology and medicine.[71] For example, in 1986, the U.S. National Institutes of Health started to require the inclusion of women as study objects in applications for medical testing and research.[72]

Interestingly, Altmann was hesitant to focus on mothers and infants for her dissertation project—due to strong expectations of women to study females—but then she increasingly realized that mothers and infants were the most interesting individuals that had the greatest evolutionary impact.[73] At the time, the focus of interest in behavioral ecology was male reproduction, females were considered to be of little interest.[74] Her studies provided detailed analyses of female variation in maternal behavior and the ensuing consequences and showed that evolutionary biologists had overrated the variation in male reproductive success and underrated female variation.[75] She used the metaphor "dual career mothering" to emphasize that female baboons constantly juggle demands of nursing, eating, socializing with other adults and being wary of dangers.[76] These detailed analyses provided empirical manifestations of parental investment trade-offs and changed the way primate females were perceived.[77] Thus, her PhD thesis, *Baboon Mothers and Infants* (1980), was important to showcase a fuller picture of female life. Furthermore, Altmann convincingly demonstrated that baboon society was not as male-focused as earlier studies and models had made out.[78]

---

70 Ibid.: 307.

71 Fedigan, 2001: 49.

72 Schiebinger, 1999: 115.

73 Haraway, 1989: 312.

74 Alberts & Silk, 2013: 198.

75 Haraway, 1989: 312.

76 Ibid.: 313.

77 Alberts & Silk, 2013: 198.

78 Strum & Fedigan, 2000: 23.

## Changing Language

As a result of the critique of stereotypical description of females, terminology use has changed in primatology—away from coy and passive females toward acknowledging female agency. Linda Fedigan calls attention to two major biases in terminology—androcentrism and a preference for metaphors of combat.[79] Feminist primatologists have been actively contributing to this shift, for example, Sarah Hrdy criticized the depiction of females as coy and passive, and Jeanne Altmann as editor for *Animal Behaviour* (1978–1983) was engaged in reforming the terminology, negotiating with authors to not use the term "rape" about animals.[80]

In scientific reports, there used to be two ways to characterize female sexuality: attractiveness (whether or not males were attracted to them) and receptivity (whether they accepted male advances)—both describing females as passive.[81] However, already in 1976, the American ethologist Frank Beach (founder of the research field behavioral endocrinology), noted that primate females in estrous seek and initiate copulations and named it proceptivity—a commonly used term since.[82]

Connected to the shift in perceptions from passive to active females is also the shift away from the theory of behaviorism (assuming that animals react without mental reflection) to the recognition of animal cognition—leading to new terminology which granted animals agency, such as social strategies, primate politics and peacemaking.[83]

The 1970s brought with it an awareness of rebounding anthropomorphism among ethologists and primatologists,[84] what media researcher Hillevi Ganetz calls a cultural boomerang, when "Nature is first culturalised according to human norms, and then this version of nature is used to naturalise the same norms".[85]

---

79 Fedigan, 2001: 56.

80 Haraway, 1989: 310.

81 Fedigan, 2001: 57.

82 Ibid.: 57.

83 Ibid.: 57f.

84 Ibid.: 53.

85 Ganetz, 2004: 209.

## Primate Models for Human Evolution

One example of a cultural boomerang is the use of primates to model early human evolution. Baboons were initially studied as such a model because they had many features in common with prehistoric humans—living in large mixed-sex groups and living on the savannah (rather than in trees), and eating a variety of foods.[86] Furthermore, because they dwell on the ground, they are also easier to watch than tree-living species (such as the howlers which Ray Carpenter studied) and came to be the first species to be extensively studied in the field. Irven DeVore and Sherwood Washburn studied common baboons, *Papio cynocephalus*, in East Africa from 1959 to 1960 and constructed the first model of human evolution based on baboons.[87] They suggested that the greater risk of predation ensued by living on the ground led to a social system in which males organized themselves with "army-like" cooperation and dominance hierarchies to protect females and young—mirroring expectations of sex roles in western society with aggressive men and submissive women.[88]

Critique of the "baboonization" of human evolution emerged from several perspectives—anthropological studies of human hunter-gatherer societies showed that 60–80 percent of their diets consisted of plants gathered primarily by women; primate studies showed that females are mostly able to take care of themselves and their offspring with little assistance from males; since recent human forest burning has forced baboons to move from forests to the savannah, it is uncertain whether baboon society is specially adapted to savannah life; later studies did not find baboons as hierarchical, aggressive and male-dominated as the first studies by DeVore and Washburn suggested; furthermore, among primates in general, a mixture of plant and animal foods is common.[89]

Later, the chimpanzee, as the ape most closely related to humans, became the new model species for early human evolution, and was

---

[86] Rees, 2009: 55.

[87] Fedigan, 1986: 39.

[88] Ibid.: 39.

[89] Fedigan, 1982/1992: 311f.

typically described as male-bonded, aggressive and male-dominated,[90] yet another model of male-dominated human evolution.

The discovery of the bonobo, which is equally closely related to humans as the chimpanzee—notably peaceful, typically sexually active within as well as between sexes, and often forming coalitions among unrelated females (which is unusual among primates)—has problematized the chimpanzee model of human evolution and provides alternative visions of what early human society might have been like.[91]

Reviewing different models of human evolution, Linda Fedigan points out that the existence of many different models based on similar information demonstrates the interpretative and cultural aspect involved in modeling early human evolution—these models reflect how we see ourselves. Fedigan asserts that currently it is recognized that one species alone will not suffice as the model for early human evolution.[92] As more of the 200 primate species have been studied, the revelation of great diversity in social systems and relationships between the sexes shows that the model species chosen to study greatly influences perceptions about those relations.[93]

## Female Orgasm

The history of the female turn also concerns shifting perceptions about female sexuality. Sarah Hrdy in her *The Woman that Never Evolved*[94] (1981) brought attention to the research about orgasm in female primates and criticized that researchers during the 1960s–1970s hypothesized that the female orgasm was a by-product of evolution of male orgasms. Likewise, feminist philosopher Nancy Tuana traces the epistemology of ignorance about female orgasms, pointing to the politics/societal values of the non-feminist accounts in evolutionary debates on orgasms.[95] She calls attention to that such accounts reduce female sexual-

[90] Parish & De Waal, 2000: 97.
[91] Ibid.: 97.
[92] Fedigan, 2001: 52.
[93] Ibid.: 52.
[94] Hrdy, 1981/1999.
[95] Tuana, 2004: 219ff; see also Lloyd, 1993.

ity to reproduction and that, until recently, most theorists considered female orgasms to be uniquely human, although male primates were acknowledged to show orgasmic responses during ejaculation.[96] American anthropologist Donald Symons in his influential *The Evolution of Human Sexuality*[97] (1979) reviewed studies on female orgasms in primates—including documentation of vaginal contractions, frowning and round-mouthed stare expression and vocalizations after clitoral and vaginal stimulations—and concluded that there was no compelling evidence that nonhuman female mammals experience orgasms.[98] In Tuana's analysis of Symons' argument, it is obvious that ignorance is not merely a lack of knowledge, it is constructed through specific exclusion of evidence—Symons only counted orgasms in heterosexual copulations in "natural" settings, dismissing observations from laboratories and zoos. Tuana concludes: "why the decades of denial of orgasm to our primate sisters in the face of their embodied pleasures? What is the logic of this epistemology of knowledge-ignorance? The desire to make the human female orgasm unique was linked to the desire to argue for the so-called pair-bond, that is, monogamous heterosexual coupling—the family values script".[99] Evolutionary theorists imagined that orgasm evolved in human females due to sexual selection to facilitate the pair-bond. The evidence from female primates' orgasms contradicts this origin story and human female multi-orgasmic capacity also does not fit. Furthermore, the oft-told story about promiscuous men and monogamous women—the hypothetical dramatic differences in number of offspring between men and women (thousands versus about 20) are not supported by the facts. Random mating with several women, men's limited sperm supplies, counting in sperm competition from other men combined with women's fertility limited to a few days a month—makes the fertilization probability small in each case and the realistic sexual difference in reproductive potential much smaller.[100]

---

[96] Tuana, 2004: 221.

[97] Symons, 1979.

[98] Tuana, 2004: 221.

[99] Ibid.: 222.

[100] Ibid.: 223.

Thus ignorance about female orgasm in primates was sustained through a narrative of evolution of the uniquely human heterosexual family—and cultivation of ignorance by discounting empirical evidence from laboratory and zoo studies. But feminist primatologists such as Sarah Hrdy advanced liberatory knowledges about female orgasms. Just as the women's movement has redefined female sexuality to not equal merely reproduction—primatologists have made visible sexual behavior between the same sex (e.g., in macaques[101] and bonobos[102]) and for other functions than reproduction (relieving social tension, social bonding and pleasure[103]).

## Multiple Feminist Strategies

Haraway showed how these feminist primatologists—in multiple ways and with different, sometimes opposing, strategies—engaged in contests to change the field of primatology toward examining females and their active strategies.[104] While Jeanne Altmann disemphasized high drama and foregrounded everyday behavior to highlight the role of females in evolutionary biology, Sarah Hrdy, on the other hand, worked from within the discourse of sociobiology and high drama of sexual conflict to emphasize the sexual politics of female primates. In Nancy Tuana's words—these feminist primatologists undermined the production of ignorance about females' lives, beings and strategies by creating liberatory knowledges about female animals as well as humans.[105]

Following on this multifaceted history of shifting perceptions about females in the field, I now turn to the particular accounts of Sarah Hrdy and Meredith Small, detailing their work and personal experiences of contributing to the female turn.

[101] Small, 1993.

[102] Parish, 1994.

[103] Tuana, 2004: 219ff; see also Bagemihl, 1999.

[104] Haraway, 1989: 350.

[105] Tuana, 2006: 1f.

## Sarah Blaffer Hrdy: Feminist Pioneer in Primatology

Primatologist Sarah Blaffer Hrdy is Professor Emerita at the Department of Anthropology, University of California Davis, and has pioneered feminist interventions in primatology through forefronting active females and questioning stereotypical perceptions about "coy" and passive females in sexual selection.[106] When I interviewed her at her farm in California, she commented on my colleagues' and my review and critique of genital evolution research,[107] she fully agreed with our analysis, but we had omitted the research on clitorises—an engaged and to-the-point criticism from one feminist biologist to another.

Hrdy grew up in Texas, in what she describes as "a deeply patriarchal, very segregated, racist part of this country" and earned her PhD in anthropology at Harvard University in 1975. She recalls her time as a cultural anthropology undergraduate student at Radcliffe College, which had started out as a women's coordinate institution to the male-only Harvard University[108] but by then had become the women's part of Harvard. There she participated in one of the first courses in primatology in the U.S., taught by Irven DeVore. During this course, she learnt that male Hanuman langur monkeys kill infants of their own species. The prevailing theoretical paradigm interpreted individual behavior as fulfilling the needs of the group, therefore infanticide by males was thought of as a sort of population regulation exhibited in high population densities. Hrdy decided to investigate it, and thus became a graduate student at Harvard supervised by DeVore, she started in the middle of the year and in the first summer—with some funds from her mother and almost no training in how to conduct field studies—went to study langurs in India. At Harvard, she also had ant expert and founder of sociobiology Edward O. Wilson and (to become) renowned evolutionary biologist Robert Trivers on her thesis committee, supervising her PhD studies. It was at the time and place of the emergence of the new field of

[106] Hrdy, 1981/1999, 1986.
[107] Ah-King et al., 2014.
[108] Radcliffe College is since 1999 fully integrated with Harvard University.

sociobiology—the study of social behavior from an evolutionary perspective—which was rather uncontroversial when applied to animals, but very controversial for its suggested application to human behavior.[109] Hrdy says that she was accepted in the mainstream sociobiological circles, even though she was a woman, in a way that she thinks she wouldn't have been if she had entered the field writing what she did later. For example, Trivers accepted the hypothesis that infanticide could be a sexually selected male strategy, but he was initially skeptical to the idea that females would mate with multiple males to avoid infanticide. However, he did come around and accepted that idea too. Hrdy's work developed at a point in time when evolutionary biology shifted from group selection (the notion that selection acts on whole groups of organisms, for example as selection "for the good of the species") to individual selection (the notion that selection acts on individuals and lead to individual adaptation), but as we shall see, she has later become more interested in cooperative than self-serving behavior, for example shared infant care (Fig. 2.1).

## Infanticide as a Sexually Selected Reproductive Strategy

Early modern western primatology was influenced by prevalent social theory (as a majority of primatologists had their background in social sciences) supposing every individual had a role to play in ensuring the survival of the society.[110] Hrdy went to the field in Northwestern India to investigate the population regulation hypothesis, and contrary to expectations, she found that males were tolerant to infants even in crowded habitats. It was only the males that took over a breeding group that attacked infants—and only unweaned infants, and furthermore only infants of females they had not previously mated with. In contrast to the population regulation hypothesis, Hrdy suggested that male infanticide is a sexually selected strategy.[111] She reasoned that males by killing young, unweaned infants can increase their reproductive success, because the

[109] Segerstråle, 2000.

[110] Hrdy, 1977a: 41.

[111] Hrdy, 1977a.

**Fig. 2.1** Primatologist Sarah Blaffer Hrdy at her farm in California, USA

infants' mothers become receptive earlier than if the infants had survived. Furthermore, she suggested that females have counter-strategies, they seek out and invite copulations with multiple males to confuse paternity and also show estrous behavior during pregnancy when a new male takes over the troop.[112] Her article "Male-male competition and infanticide

[112] Hrdy, 1974: 50ff.

among the langurs (*Presbytis entellus*[113]) of Abu, Rajasthan" was published in *Folia Primatologica* in 1974, followed by a further generalization of the adaptive infanticide hypothesis to other primate species,[114] and her 1977 book *Langurs of Abu: Female and Male Strategies of Reproduction*. Her suggestion ignited a controversy about the interpretation of infanticide in primates.[115] Hrdy says that within the field of evolutionary biology, the controversy ceased as further evidence of infanticide accumulated from other primate species, but within anthropology, the controversy continued. Science historian Amanda Rees has studied this controversy and ascertains that the controversy is unresolvable due to the different scientific communities (within primatology) having established their own standards for making judgments on validity,[116] where the proponents of the sexual selection hypothesis focused on whether circumstantial evidence fulfilled predictions of the hypothesis, and the opponents emphasized the selection of aggression in general as a possible explanation and that therefore the level of infanticide need to be compared to general patterns of morbidity.[117] Furthermore, Rees suggests that part of the reason for the long-lasting opposition to the sexual selection hypothesis, especially in the US where primatology was institutionally positioned as part of anthropology, was its implications for interpretations about humans.[118]

The infanticide controversy was a struggle for Hrdy, and she says that for a long time it made it hard for her to get an employment in anthropology. Just after she received her PhD, during the sociobiology controversy, the only Anthropology Department in the U.S. that was looking for a sociobiologist sought to recruit her, but as they customary approached her advisor—who just noted that she was married and proposed another male (married) sociobiology primatologist—she did not get the job.[119] A

[113] The Latin name of Hanuman langurs has since 1974 changed from *Presbytis entellus* to *Semnopithecus entellus*.

[114] Hrdy, 1977a, 1977b.

[115] See Rees, 2009.

[116] Rees, 2009: 217.

[117] Ibid.: Chap. 6.

[118] Ibid.: 179f.

[119] Hrdy, 2010: 367.

clear example of sex discrimination, but she did not even know about it until several years later. Nevertheless, Hrdy thinks that the controversy helped her career as a woman in a very male-dominated field—if it was not for the attention of the controversy, her hypothesis might have been co-opted as many pioneering ideas had been for other women in science. Due to the controversy, she was given the honor of having suggested it, and no one could take it away from her.[120] Thus, she was elected a member of the National Academy of Sciences (1990) at an unusually young age. After temporary positions at Rice University (Texas), as an unpaid research associate at the Peabody Museum (Massachussetts) as well as working as a volunteer at her children's daycare centre, in 1984 she became a full professor at the University of California-Davis, benefiting from a special program for outstanding women.[121]

## Empathy with Females of Other Species

Several researchers have suggested that women in science have brought with them empathy for their female study objects and thereby forwarded gender-reflective approaches in science. Sarah Hrdy says that coming up with the sexual selection hypothesis for infanticide by males "raised all these questions for me as a woman—wait a minute, what are females doing about it?" Although she did not yet have children, she recalls that she identified with the langur females and it made her think about infanticide from the perspective of the female. Hrdy also talks about her own experience of a monthly period of increased libido as an experience that she brings to her study of primates—a perspective that male researchers perhaps don't have. She describes herself as politically naive when she started studying primates, it was only later that she came to identify as a feminist. Feeling empathy with the female langurs, she started asking questions that her male colleagues had not been asking, questions that they might consider uninteresting, or that they thought they already knew the answers to. She paid attention to the curious observation of a

[120] Sarah Hrdy interview 2017.

[121] Hrdy, 2010: 367.

female leaving her group and soliciting copulations—why would she do that? Other researchers might have dismissed it as an abnormality. She suggests that women scientists were less likely than men to identify with scientific status quo—being on the margins of the scientific community themselves, they might be more prone to go against prevailing ideas about sex roles than the professors, who were predominantly men.[122] Women were also more likely to be affected by feminist ideas, but she does not think that women and men differ in their ways of conducting science[123] (see further discussion about empathy below).

## A Male-Biased Field

Professional primatology has shifted from a male-dominated field into including an increasing number of women. Sarah Hrdy was Irven DeVore's first female graduate student, and at the time when she graduated from Radcliffe College, women undergraduates were not allowed into Harvard's Lamont library where the undergraduate textbooks were kept. Natural sciences was a men's world. Starting out in the social sciences, Hrdy found an advisor in social anthropology who was very supportive, but she says that "things went downhill, as I moved towards the sciences". There were no women professors at Harvard's Faculty of Arts and Sciences when Hrdy graduated in 1969, she had to look elsewhere for women role models and was influenced by Alison Jolly (working on lemurs in Madagascar, and working as an adjunct at the University of Sussex, UK at the time) and Jane Lancaster (studying vervet monkeys at Berkeley). She says graduate school in a male-dominated science department was a hostile environment for women with professional aspirations, although there were rare exceptions of supportive male seniors.[124] Her empathy with female langurs, her experiences as a woman in a male-dominated field, as well as her readings of feminist critiques of

[122] Hrdy, 1981/1999: xviii.

[123] Ibid.: xvii.

[124] Hrdy, 1986: 142.

androcentrism in science inspired her to challenge conventional assumptions in science.[125]

## Male-Biased Responses

Part of the production of ignorance about females occurs in the reception of new findings in the scientific community. In *Langurs of Abu* (1977), Hrdy had described infanticide by males as well as female-female competition and she devoted equal effort to analyzing male and female strategies, and "almost equal number of words to male and female perspectives". Despite this equal treatment by Hrdy, the general response to the book was focusing on the male perspective,[126] as did for example British zoologist Tim Clutton-Brock in his review of the book in *Nature*.[127] Hrdy confronted Clutton-Brock and questioned this male bias and he replied that it is "just so much harder to study female than male reproductive success". Elsewhere, Clutton-Brock explains "there are reasons why it can be harder to collect detailed data on female breeding strategies than on those of males—female mammals are often less visible than males and their longer breeding lifespans and reduced variance in seasonal success complicate the study of breeding strategies".[128] However, this discussion took place before DNA analyses of parentage were available, it was clear how many offspring a female had given birth to, but the way primatologists were estimating reproductive success in males was through "counting cops"—counting copulations. At the time, female reproductive success was considered nonsensical, recalls Hrdy. Against available evidence, theoretical focus was on male reproductive success and female reproductive success was ignored, as Linda Fedigan also showed in her 1983 review (see above). Here, production of ignorance takes place in the reception of Hrdy's studies in the scientific community, a reception in line with the prevailing androcentric theoretical framework.

---

[125] Hrdy, 1986: 140.
[126] Hrdy, 2010: 362.
[127] Clutton-Brock, 1978.
[128] Clutton-Brock, 1983: 716.

Hrdy mentions another example, the reactions to her paper on hierarchical relations among female langurs published in *Science* in 1976.[129] This paper was pioneering in western dominance studies, suggesting a new system of dominance hierarchy based on female reproductive value (potential future reproduction—younger females have higher reproductive value than older), rather than the already known dominance systems determined by size, age or genealogy. Yet, the response to the paper was all about what the males were doing. Hrdy suggests that people were not very interested in selection pressures on females and causes of female reproductive success variation—all that was under the radar. Hence, the assumption that females vary little in reproductive success between themselves and that sexual selection mainly concerned males contributed to the production of ignorance about females.

## Criticizing Ignorance of Females in Evolutionary Biology

Time and again, identifying gender bias in science has motivated feminist scientists to revise perceptions about females in their fields. Frustration with the male-focused responses to *Langurs of Abu* motivated Hrdy to focus on females in her next book, *The Woman That Never Evolved* (1981). There, she criticizes sexual selection theory's assumptions about coy and passive females which were contradicted by recent primate studies showing female initiations of mating with multiple males, mid-cycle wanderlust, and evidence of orgasms. She emphasizes the importance of competition between females—the book is an exposition of ignorance about females and an argument for revision of evolutionary theory applied to females, and for paying attention to females' active sexual strategies in primates as well as in humans.

"When *The Woman That Never Evolved* first appeared, in 1981, I feared it would terminate a promising academic career—namely, my own,"[130]

[129] Hrdy & Hrdy, 1976.

[130] Hrdy, 1981/1999: xiii.

writes Hrdy, and explains that at that time there was no overlap between feminism and evolutionary biology, feminists raged at sociobiology and sociobiologists were puzzled by feminist arguments.[131] Despite this seemingly unreconcilability between feminists and sociobiologists, *The Woman That Never Evolved* was positively reviewed in the radical feminist newsletter *Off our backs* and the fierce criticism from feminists that Hrdy anticipated failed to arrive. The book was met with some skepticism[132] but was also widely read among evolutionary biologists, leading the way to research with gender-reflective perspectives—she went from a critic from the margins in 1981 to being part in the stampede of putting females at center stage by the 1990s.[133] Support from prominent biologists George C. Williams,[134] John Maynard Smith and William Eberhard to reinforce discussions about gender bias in sexual selection research stands out in her memory.[135]

In the same line of feminist criticism, Hrdy published "Empathy, polyandry and the myth of the coy female" in 1986 in *Feminist approaches to science* edited by feminist neurophysiologist Ruth Bleier. Hrdy notes that the chapter was overlooked at the time of its publication, but got noticed when it was republished in *Conceptual issues in Evolutionary Biology* edited by Elliott Sober in 1994. In this chapter, Hrdy points out the basis for the stereotypic assumptions of coy, highly discriminating and passive females—the assumptions that male contribution to offspring was relatively small, that there was little variation among females in reproductive success compared to the great variation among males and that fertilization was the only reason for females to mate.[136] She notes that although these assumptions are appropriate in some circumstances, they are far from universal and that their uncritical acceptance has hampered the development of knowledge. Hrdy reviews shifts in perceptions and theoretizations about females as well as the important role of women

---

[131] Ibid.: viiif.

[132] For example, by evolutionary psychologist Donald Symons (1982).

[133] Hrdy, 1981/1999: xvii.

[134] Hrdy & Williams, 1983.

[135] Ibid.: xviif.

[136] Hrdy, 1986: 119.

researchers for these shifts. She concludes with antidotes to biases—common sense in methodology (it was no longer accepted to draw conclusions about primate societies from studies of only one sex), recognition of sources of biases, to include a broad spectrum of observers and being self-reflective and open to criticism.[137] I interpret Hrdy's strategy as mainly following what science philosopher Sandra Harding has described as a feminist empiricism, namely to urge for increased awareness of gender bias in combination with following rigorous scientific empiricist methods,[138] although she also includes recognition of different observers' partial perspectives.

## Challenging Assumptions About Mothers and Care

In addition to questioning females as coy and passive, Hrdy also questioned assumptions about self-sacrificing and ever-caring mothers.

In 1977, Hrdy's first child was born and as she was influenced by attachment theory, she says she felt like she had to be like a chimpanzee mother and stay in contact with her baby all the time. However, she soon realized that this was in conflict with her other desires, such as doing science, and she felt very ambivalent about it. In her book *Mother Nature* (2000), she scrutinizes the common assumptions that mothers are naturally self-sacrificing, nurturing and caring, and shows that females may abandon and even kill their offspring if they do not have resources enough to rear them. In other words, mothers too are active strategists. She developed the cooperative breeding hypothesis, that humans could not have evolved to have so dependent babies unless mothers had a lot of help from others for provision. In the subsequent *Mothers and others: the evolutionary origins of mutual understanding* (2009)[139] she suggests that the rearing conditions in cooperatively breeding humans lead to more sustained development and selected for a greater capacity for understanding others—in order for babies to be able to distinguish among the larger circle of carers who would help and who would hurt. She also suggests

[137] Ibid.: 141.

[138] Harding, 2006: 237f.

[139] Hrdy, 2009.

that cooperative breeding led children to develop adaptations to elicit care. Hrdy presently continues her line of research about human evolution by exploring the nurturing capacity of males.

## A Feminist Pioneer

As a feminist pioneer in primatology, Hrdy has persistently questioned assumptions about females as coy and passive, self-sacrificing mothers and female (non)sexuality. First positioned at Harvard, at the heart of emerging sociobiology, during the shift from group selection to the individual selection paradigm, she formulated the hypothesis of sexually selected infanticide by males. In addition, she explored langur females' responses to infanticide—and suggested that their behavior to seek out and mate with many different males was a counter-strategy to infanticide by males, thereby confusing paternity certainty and increasing the chances of their offspring's survival. Hrdy calls herself a feminist sociobiologist, and with sociobiology being criticized by many feminists, this position has given rise to contradictory treatment of her among academics[140] and in the feminist science literature.[141]

Thus, Sarah Hrdy's situated knowledges—formed through her empathy with female primates, her study species, sociobiological theory, feminist insights and sexed/gendered lived experiences of being a woman and mother in a male-dominated science—influenced her understanding of sex and sexual selection. Her account gives several examples of ignorance production about females, the androcentrism in sexual selection that she criticized, the prevalent assumptions about invariable females, the skepticism toward female multiple mating as a strategy to prevent infanticide, and the scientific community's male-focused reception of her studies which largely ignored her thorough attention to selection on females. Her struggles to make evolutionary biologists pay attention to female behavior, in combination with her successful efforts to write for the

---

[140] Van Schaik & Silk, 2013: 201.

[141] Fedigan, 2001: 60.

general public have contributed significantly to shifting perceptions about females in the field and to steer attention to selection in females.

## Meredith Small: Female Primates and Their Power Relations

Biological anthropologist Meredith Small is a Professor Emerita of Anthropology at Cornell University, with a PhD from University of California, Davis in 1980. She has made a career out of focusing on female primates before she turned a popular science author. She entered the field of primatology some years later than Sarah Hrdy and her account differs from Hrdy's, especially noticeable is Small's experience of a lack of resistance to her research focus on females.

### The Influence of the Study Species

"I was called that I wear macaque glasses"[142] explains Small, as I interview her in her visiting writing residence in Venice, Italy. Like other American anthropologists, she studied primates with the aim to shed light on human evolution. Thus, influenced by her studies of macaques, she views humans as "chatty, socially interactive and keeping track of everything", and she contrasts her understanding to her colleague John Mitani's, who was working with solitary orangutans, and therefore held very different ideas about humans and primate evolution. This difference in perspectives illustrates how evolutionary researchers are influenced by their specific study species, as one aspect of forming their situated knowledges.

She recalls her first contact with primatology when she joined a course on primate behavior in 1975–1976, at a time when professional primatology was still in its cradle. The course had Irven DeVore's *Primate Behavior: Field Studies of Monkeys and Apes* for reading, which was descriptive and female primates were only written about in their role as mothers. "Hasn't anybody done anything on females without infants?" Small asked

[142] Meredith Small interview 2017.

the lecturer, who stopped to think and then answered that he didn't know. Starting with the question about females without infants, Small contacted primatologist Peter Rodman at the Primate Center in Davis, California, and was invited to study their captive rhesus macaques (*Macaca mulatta*) and bonnet macaques (*Macaca radiata*). She did not experience resistance about her choice of subject or perspective, rather the primatologists she encountered encouraged her to fill out the knowledge gap about female primates. Furthermore, by then, there were supporting feminist pioneers in primatology. The group working at the Primate Center in Davis was a diverse assembly of primatologists with backgrounds in animal behavior, psychology and anthropology, and it was as Donna Haraway points out, one of the institutions that became a nodal point for reinterpreting females, training women scientists and assembling a network of women primatologists.[143] Thus, many of Small's colleagues were women, including Hrdy, and she found the atmosphere at the Primate Center to be amicable and supportive. Small notes that she was the only one studying female choice at Davis at that time. However, all the male primatologists did study females as well and were engaged in gender-reflective approaches—she says the men were living through the sexual revolution themselves and were influenced by what women around them were doing. She describes it as a wave of studies on female primates, from the late 1970s that lasted about ten years. In addition, her colleague Jeff Kurland—who had studied Japanese macaques and been involved with Japanese primatologists who had already identified female power and social organization around matrilines (where maternal lineage is important for social status)—"at the drop of a hat would talk about female power in these animals". The influence of Japanese primatology also served to shift perceptions about females, although, as we shall see later in this chapter, their findings were initially overlooked and ignored by western primatologists (Fig. 2.2).

[143] Haraway, 1989: 302.

Fig. 2.2 Primatologist Meredith Small in Venice, Italy

## Questioning Prevailing Assumptions About Choosy Females

After her PhD, Meredith Small went to study Barbary macaques in a captive population in France. She describes this species' power relationships as "females basically own those groups". Her observations of sexual behavior in female Barbary macaques challenged prevalent views on sexual selection—they initiate mating with multiple males in rapid succession, without gaining increased male care or protection.[144] These observations contradicted prevalent predictions that females with large parental investment perform a careful and selective choice of mates. In 1989, Small reviewed the evidence for female choice among primates,[145] and showed that females in 20 species of primates have preferences for males based on familiarity, novelty or status. Later, she wrote a popular science book *Female Choices: Sexual Behavior in Female Primates*,[146] emphasizing that females (including humans) have diverse active sexual strategies—preferring unfamiliar males, mating with several males or mating indiscriminately.

In an endeavor to both exhibit the increase in studies of female primates and to acknowledge the important contributions of women researchers, Small edited the book *Female Primates: Studies by Women Primatologists* (1984). Donna Haraway's analysis of this volume calls attention to the ways in which these studies have disrupted and restructured primatology and at the same time that the primatologists describe this progress in terms of feminist empiricism (adherence to scientific methodology to correct bias)—without challenging narratives of how science is made or criticizing political social relations within primatology.[147] (See discussion on the influence of women, feminism and gender-reflective approaches in primatology later in this chapter.)

## Experiences of Resistance

Although Small did not meet resistance within primatology, she experienced opposition in paleontology and in anthropology. In the very male-dominated field of paleontology (the study of ancient life through fossils), in which Small did her Master's degree, women's participation in fieldwork was questioned. But the opposition she experienced as most severe was the anti-evolutionist resistance she later encountered at the anthropology department at Cornell University. She says what was most

---

[144] Small, 1990: 267.

[145] Small, 1989.

[146] Small, 1993.

[147] Haraway, 1989: 290ff.

challenging to her was "to be faced with anthropologists who seriously straight out say evolution has nothing to do with human behavior". The sociobiology controversy led to a split between social and biological anthropology in many anthropology departments throughout the US.[148] As the only biological anthropologist in an otherwise cultural anthropology department, Small was caught in the aftermath of this controversy, she met resistance that had nothing to do with her studies focusing on females—it concerned the questions of whether biology or culture determines human behavior and whether God's creation or evolution is the valid explanation for the origin of Earth, humanity and life—that is creationism versus evolution. Small says creationism has a very strong hold in the U.S. When applying for tenure, she was asked to include "other ways of evolution" in her book manuscript on *Female Choices: Sexual Behavior in Female Primates*,[149] but she refused and managed to get tenure anyway, since the application was assessed by evaluators outside of the department.

## Small's Situated Knowledges

When reflecting on how she came to study females and their power relations, Small says that coming from a middle-class family with no money, with a working mother and being brought up to support herself financially, she started working and earning money at the age of fifteen and highly values her independence. To her, making her own money, not getting supported by her parents, and not being financially dependent on a husband—is freedom. That personal history is just in line with studying females, their independence and power, in her words "their agency, their desires, their choices, their wants, their ability, their intellectual ability—even if you're a macaque—to make a decision". Her perception of female monkeys as active strategists comes from hours of watching especially female bonnet macaques—seeing them form strong relationships, bonds between sisters, and also antagonistic relationships with each other. She

[148] See Rees, 2009: Chap. 6.

[149] Small, 1993.

notes that "male bonnet macaques are actually pretty nice", in comparison to the nasty male rhesus macaques, and that female bonnet macaques have a certain agency and power in their matrilinear groups.

Her account reiterates the androcentrism of western primatology, and how focusing on females and pursuing gender-reflective approaches became a way of unmaking ignorance about females.

Small's situated knowledges are formed by her feminist insights, her emphasis on economic independence, agency, focus on females in themselves (independent of their role as mothers), her study species (especially Bonnet macaques in which females have relative freedom from male harassment) and evolutionary theory on social behavior (sociobiology)—led her to question assumptions about passive and sexually reserved females. Her research and popular science writing, even cited in the UN decade for Women's report (1985)[150], contributed to the illumination of female choice and female agency in primatology.

## Overlooked Females in Anthropology: Man the Hunter and Woman the Gatherer

Androcentrism was not confined to primate studies, the subsequent suggestions of the Man-the-hunter and Woman-the-gatherer hypotheses in evolutionary anthropology are another example of gender bias and male precedence. Since primatology was initially taught mainly by anthropologists in the US, the history of primatology and anthropology there is intimately connected. Models of human evolution suggested in the 1960s shared the common theme of "man the hunter"—"with a strong focus on one distinguishing human trait: the pursuit, killing, and eating of animals with the use of tools".[151] In 1966, anthropologists Irven DeVore and Richard Lee organized a "Man the Hunter" symposium at the University of Chicago, resulting in an edited book (1968), whereof the most influential paper was Sherwood Washburn and Chet Lancaster's "Evolution of Hunting" (1968). They argued that hunting required qualities in human

[150] Haraway, 1989: 287.

[151] Fedigan, 1986: 32.

behavior that distinguish them from other primates, namely "bipedalism, elaborate tool kits, development of language, appreciation of beauty, male aggressiveness and pleasure in killing, division of labor, the nuclear monogamous family, loss of female estrus, the invention of incest taboos, and bonding between males".[152]

In 1970, Sally Linton (later Slocum) presented a paper on "Woman the Gatherer: Male bias in Anthropology"[153] at the American Anthropology Association meeting in San Diego, in which she criticized the male bias of the "Man the hunter" hypothesis and drew on different sources of evidence to develop a model on early female hominids gathering, carrying and sharing food with their children. She saw this as a logical extension of the female primate-infant bond and argued that male hunting of large game would have come later. Her model was the first to portray autonomous females as an important factor in how primates evolved to become humans.[154] Linton's critique was an important inspiration for many women anthropologists, for example Adrienne Zihlman,[155] who further developed the idea of "woman the gatherer" (during 1979–1986), arguing that bipedal walk and slings enabled women to carry their babies as well as gather plant foods from larger areas and bring them back to safer areas to share with their children. In this origin story, women were the innovators and providers.[156]

The suggestion of woman the gatherer did not get much attention among paleoanthropologists, it was either overlooked or co-opted, and the co-opting was expressed in two ways: either the theory was seen as additional to man the hunter, resulting in the same pair-bonding and sexual division of labor as for "Man the hunter", or admitting gathering tools as important for human evolution but reassigning them to males, thus still focusing on nuclear families.[157]

The latter strategy was used in Owen Lovejoy's very influential article "The Origin of Man" (1981) in *Science*—representing contemporary

[152] Ibid.: 32f.

[153] Linton, 1974.

[154] Fedigan, 1986: 34.

[155] Haraway, 1989: 334.

[156] Fedigan, 1986: 35.

[157] Haraway, 1989: 345.

orthodoxy about human evolution.[158] He suggested that humans evolved longer lifespans, slower rate of growth and yet higher reproductive rate than apes because humans evolved pair-bonding, division of labor and men contributed to their partner's reproductive success through provisioning by hunting—thus a version of the familiar "women dependent on men" origin story.[159] Hence, the scientific community's response produced ignorance about females, by reinterpreting suggestions of autonomous and innovative women into the prevalent androcentric story of women being dependent on men.

Linda Fedigan observes that feminist thinking blossomed at different times in different subfields of anthropology—in sociocultural anthropology in the 1960s and 1970s, in primate studies in the 1970s and 1980s, in archeology in the late 1980s and 1990s, and hardly at all in paleontology.[160] In sociocultural anthropology, there was a female turn in understanding women's lives when women anthropologists started studying women. Sarah Hrdy recounts that it was because women anthropologists in the field got access to talk to women, something men could not do in many cultures. I will get back to the role of women, feminism and gender-reflective approaches for the development of primatology.

## Japanese Primatology: Early Recognition of Female Power

Japanese primatology developed independently from western primatology and has a different story of perceptions about females. Japanese primatology started with studies of Japanese macaques (Fig. 2.3) in 1948,[161] organized by animal ecologist Kinji Imanishi, founder of the influential Kyoto school.[162] Japanese primatologists early on made discoveries about

[158] Fedigan, 1986: 37.
[159] Ibid.: 36f.
[160] Fedigan, 2000: 513.
[161] Rees, 2009: 28.
[162] Takasaki, 2000: 151.

**Fig. 2.3** Photographs of study species. (**A**) Japanese macaques, *Macaca fuscata*, studied by Japanese primatologists such as Syunzo Kawamura. Photo by Asteiner Creative CommonsAttribution 3.0 Unported. (**B**) Hanuman langur, *Semnopithecus dussumieri*, female with infant, the species studied by Sarah Hrdy. Photo by Yann Creative Commons Attribution-Share Alike 4.0 International, 3.0 Unported, 2.5 Generic, 2.0 Generic and 1.0 Generic license. (**C**) Female Barbary macaque, *Macaca sylvanus* and her infant, one of Meredith Small's study species. Photo by Hamza-sia Creative Commons Attribution-Share Alike 3.0 Unported

primate social organization that their western counterparts missed,[163] from the perspective in this study of the female turn, the differences in perceptions about females are especially interesting. For example, Japanese primatologists found that among macaques, relations among female kin determine dominance hierarchies (1958 versus 1960s by Donald Sade, see above), and that maintaining the alpha male position is not solely dependent on physical strength, but on support from the core female members of a troop (1961 compared to 1982 in western primatology[164]). Anthropologist Pamela Asquith demonstrates how Japanese primatologists made these discoveries due to their holistic approach to studying primate societies, their attitude that a researcher needs long-term experience to study primates, and their anthropomorphic method.[165]

## A Sociological Perspective on Primates and the Human-Animal Continuum

Like the western primatologists, Kinji Imanishi was interested in understanding human evolution, but his focus was on the origin of society—and in his view, society could not be reduced to the individuals it consisted of or their relationships[166]—but needed to consider the "whole life" of the "whole species" or "species society."[167] With the species society in mind, it was important to gather detailed information on variation in behavior and group structure within a species over long periods of time.[168] In contrast to the western evolutionary science, Imanishi did not view nature as competitive, but inherently harmonious in which each species has its own complementary role in the ecosystem.[169]

---

[163] Asquith, 1996.
[164] Ibid.: 252.
[165] Ibid.: 239ff.
[166] de Waal, 2003: 293.
[167] Asquith, 2000: 167.
[168] Ibid.: 167.
[169] de Waal, 2003: 293.

The Japanese idea (from Confusianism and Buddhism) of a continuum between humans and animals means that, compared to western scientists, they do not see the observer/object and nature/culture as dichotomies.[170] While western scientists aim for a neutral objective relation to the animals they observe, and therefore consider studies of wild unprovisioned animals the most reliable, early Japanese primatologists provided free-ranging animals with food in order to better observe them, and this was not viewed as a violation of their natural status.[171] Primatologists in Japan use what has often been described as sociological, ethnographic or anthropomorphic methods in interpreting primate culture.[172]

Also, practices coded as empathetic or intuitive in the west are not deemed unscientific or feminine in Japanese culture, and thus did not hinder primatologists from using such approaches, even though this is a male-dominated field in Japan.[173] On the contrary, Kinji Imanishi emphasized identification with the animals.[174] Japanese primatologists use what they call an anthropomorphic method, since they assume that monkeys have some sort of mind, they think that an empathetic approach is required to understand primate societies.[175] In Dutch primatologist Frans de Waal's words: "The study of animal behavior in Japan has never been contaminated by feelings of superiority or an aversion to acknowledging humanlike characteristics in animals."[176] In contrast, the assumption that monkeys are not different from humans motivated Japanese primatologists to chart primate genealogies over multiple generations, discover kinship networks, paternal care, cultural transmission and female power relations.[177] Already in the 1950s, they found out that dominance was mostly dependent on female kin relations.[178]

[170] Haraway, 1989: 246ff.

[171] Ibid.: 248.

[172] Ibid.: 249.

[173] Ibid.: 246.

[174] Ibid.: 248.

[175] Ibid.: 252.

[176] de Waal, 2003: 294.

[177] Takasaki, 2000: 153ff.

[178] Haraway, 1989: 253.

Importantly, the idea of a continuum between humans and animals does not preclude hierarchies within the unity, but rather insists on dominance and subordination—Japanese studies of primates often concern status, leadership and social change.[179] Another important influence is the Shinto tradition which, with its powerful female deities, emphasizes the power of mothers in social life,[180] and thereby enables perceptions of powerful and active females. Yet, even in Japanese primatology the female-centered primate society was discussed mainly in terms of understanding especially male life patterns and explaining conservation of social systems, Haraway notes that all the above features—a holistic approach, empathic method, lack of human-animal boundary, observation of female-centered society—"are all perfectly compatible with masculinism in epistemology and male dominance in politics".[181] Thus, despite the discoveries of female dominance and agency, ignorance about females was produced to some extent also in Japanese primatology.

## Influence on International Primatology

At first, western scientists were unaware of the Japanese studies, it was a medical primatologist at Yale who in 1957 found a book with primates on the cover in a bookstore in Japan.[182] In the 1950–1960s, Japanese reports started being translated into English and becoming available to western primatologists.[183] Ray Carpenter, who was interested in social relationships, became the first western supporter of Japanese primatology.[184] However, Japanese primatologists presenting their findings in the US in 1958 were met with skepticism at their abilities to recognize so many individual monkeys.[185] Even though Japanese primatologists published their abstracts in English in their journal *Primates* since 1959,

179 Ibid.: 146f.
180 Ibid.: 247; 254.
181 Ibid.: 254.
182 Ibid.: 254.
183 Strum & Fedigan, 2000: 10.
184 de Waal, 2003: 295.
185 Ibid.: 295.

western researchers largely ignored their findings—only many years later were these findings tested with western methodology and found valid.[186] Furthermore, while details of behavioral data from Japanese studies were sometimes cited in western reports, Japanese hypotheses about the primate society were largely ignored.[187] Still, not long ago, western primatologists considered Japanese primatology to be atheoretical, too anthropomorphic and inconclusive—and the cultural divide between the two traditions continue.[188] This is an example of the cultivation of ignorance through the undermining the authority of certain knowers. However, since the 1980s the two traditions have come closer, both draw on findings from the other and there are successful cooperative projects.[189] Also, the basic tenets of Japanese primatology have been adopted by western animal behavior—to identify individuals and follow them during their life, and anthropomorphism is not considered as problematic as before.[190] de Waal explains this "silent invasion" partly by the arrogance of the native English-speaking scientists, ideas presented by non-native speakers are less likely to be taken up or are co-opted with another term which makes invisible its origin, thus producing ignorance by undermining the authority of certain knowers, and partly by the differences in articulating theory. The Japanese primatologists did not lack theory or assumptions—although that is how western scientists perceived them—but in contrast to western scientists, they are reluctant to describe their assumptions and expectations in a way that invites disagreement.[191] Still, when it comes to ideas about primate society, Japanese ideas have won out—for example, Jane Goodall's initial individualized description of the mother-offspring as the only bonded units among chimpanzees (*Pan troglodytes*), has been replaced by the Japanese discoveries of larger unit groups.[192] Japanese primatology, although largely unacknowledged, has helped western scientists to rid themselves of some of their cultural

[186] Asquith, 1996: 240.
[187] Ibid.: 249.
[188] Ibid.: 252ff.
[189] Ibid.: 254.
[190] de Waal, 2003: 295.
[191] Ibid.: 297.
[192] Ibid.: 298.

baggage colored by the traditional western dualisms of human/animal and culture/nature.[193] However, Japanese primatologists continue struggling to express what they want to convey and get their findings published in international journals.[194] Yet, despite the common disregard of Japanese influence in international primatology, primatologists seem to share an awareness that culture affects scientific discoveries.

The epistemology of primatology is culturally and historically specific,[195] but as primatologist Hiroyuki Takasaki argues, diversity in perspectives enables the illumination of topics neglected by others, "There are always some parts of the world that may be viewed better upside down".[196]

The history of Japanese versus western primatology shows the production of ignorance about females through western primatology's androcentrism and undermining the authority of Japanese researchers, which led to delays in recognizing their findings. It also shows how researchers' knowledges and ignorance are situated by their cultural context.

## Anthropomorphism and Empathy

As we have seen, Japanese primatologists use the anthropomorphic method of empathetic understanding as a scientific tool.[197] In contrast, western primatology (and animal behavior research generally) considers anthropomorphism to be a taboo. Yet, primate researchers are left to use existing terms to describe primate behavior.[198] The danger of anthropomorphism is perceived as greater for animals that are closely related to us, who look and behave like us. Linda Fedigan notes that "careful observations by scientists have convinced us that the gorilla does not beat his chest because he is angry; the vervet does not present his posterior to the face of another vervet to insult him, and the macaque does not grin

[193] Ibid.: 298.
[194] Asquith, 2000: 176ff.
[195] Haraway, 1989: 247.
[196] Takasaki, 2000: 164.
[197] Fedigan, 2001: 62.
[198] Ibid.: 56.

because she is happy",[199] demonstrating that spontaneous anthropomorphic interpretations can be misleading. The solution for western primatologists has been to operationalize language and use terms that are as void of anthropomorphizing aspects as possible, such as "open-mouth gape" rather than "threat face".

Yet, despite the taboo against anthropomorphism, Fedigan has listed a number of Euramerican primatologists that have claimed to use empathy to understand primates—Sarah Hrdy, Thelma Rowell, Pascale Sicotte, Christine Nisan, Meredith Small, Shirley Strum and George Schaller, whereof only the latter is male.[200] Fedigan notes that referring to empathy is risky for Euramerican women primatologists, both because their hard work might be dismissed as "female intuition", and because it might endanger their reputation as objective observers.[201] Fedigan herself had experienced her work being dismissed as intuition: "Having put many hours of effort into learning to identify the individual female monkeys of a large group, my ability was dismissed as being inherent in my sex, by a respected and senior male colleague."[202]

When science historian Evelyn Fox Keller described geneticist Barbara McClintock's deep understanding of corn as "a feeling for the organism",[203] it was often taken as evidence that women do research in a "feminine" way or even that McClintock was a role model for "feminist science."[204] However, what Keller meant with "a feeling for the organism" is more complex than what is generally understood—it concerns appreciation of individual differences rather than an essentialist gender projection, she did not argue that women use distinctive research methods.[205] Keller described McClintock's method as patiently and open-mindedly "hear what the material has to say to you … let it come to you".[206]

---

[199] Ibid.: 56; Male gorilla chest beating is a signal in male-male competition and mate choice (Wright et al., 2021); male vervets show their bottoms to subordinates in a display of dominance (San Diego Zoo, 2021); in rhesus macaques, the grin is a ritualized submission signal or signals fear of a dominant, versus in Tonkean macaques it signals affiliation (Sterck & Goossens, 2008).

[200] Fedigan, 2001: 62.

[201] Ibid.: 62.

[202] Fedigan, 1984: 308.

[203] Keller, 1983.

[204] Schiebinger, 1999: 6.

[205] Ibid.: 6.

[206] Keller, 1983: 198.

Science historian Londa Schiebinger points out that easy stereotypes about women and "feminine" qualities can be unnecessarily divisive, and may alienate sympathetic men who also empathize with their subjects.[207] Furthermore, Donna Haraway criticizes the assumption that women empathize with female subjects and men with male subjects and that this will cancel out bias—instead she points to the force of disruptive narratives in restructuring knowledge.[208] The insights of Japanese primatologists as well as feminist primatologists show that empathy, "a feeling for the organism", can be productive in knowledge-making.

## Primatology Politics

Donna Haraway points out that among practicing scientists, the story about the development of a discipline is conventionally told and believed to happen "inside" science, with over time less and less influence from "outside" society. This progress narrative about primatology emphasizes how early descriptive methods were replaced by more scientific theories and methods and thereby correcting earlier biases, such as male bias. This progress narrative of science history makes politics invisible by its division of influences as "inside" versus "outside" science.[209] In contrast to such progress narratives, Haraway's analysis demonstrates the politics of primatology—how power relations of race, sex and class co-constituted primatology through its development of institutions, funding, publishing and building of networks—and how current societal issues and ideologies have affected research about and interpretations of primates. In other words, the society "outside" of science is inextrinsically entwined in science throughout history.

Wild primates are found in the tropics, in Africa, Asia and Latin America, mostly in the third world. Haraway shows how the history of western primatology is "intimately a part of the history of race, sex and

[207] Schiebinger, 1999: 7.

[208] Haraway, 1989: 292–293.

[209] Haraway, 1989: 124f.

class in a world capitalist system".[210] She demonstrates that primatology in the United States has "nearly exclusively been practiced by white people, and until quite recently by white men, and is still practiced overwhelmingly by the economically priviligied"[211] and how that has influenced which questions have been prioritized. For example, in the post-World War II anxieties about social stability, primatology was dominated by "man the hunter", the mother-infant pair and theoretization about evolution of "the family."[212] After 1975, Haraway reads the "cultural logic of late capitalism"[213] in the sociobiological shift to individual strategic possibilities and animal, as well as human, behavior becoming reconceptualized as a strategic game. It is not a coincidence that perceptions about females shifted in primatology in parallel with the reworkings of women's role in society, at the same time as the United Nations declared the Decade for Women 1975–1985,[214] the passive female primate was transformed into an equally rational creature as the male.[215] Haraway shows how primatology has been influenced by societal ideologies, but also how it has been used by these ideologies to justify what is "natural", in terms of heterosexual family, sex differences and division of labor. Thus, shifting perceptions about females in primatology is associated with wider negotiations about the role of sex and sexuality in society.

## Influence of Women, Feminism and Gender-Reflective Approaches on Primatology

Many science analysists have noted that the development of a feminist transformation in North American primatology occurred at the same time as feminist movement made progress in American society.[216]

[210] Haraway, 1986: 78f.

[211] Haraway, 1989: 7.

[212] Ibid.: 126.

[213] Ibid.: 350.

[214] Ibid.: 286.

[215] Ibid.: 178.

[216] Hrdy, 1986, Haraway, 1989, Fedigan, 2001.

Although some women had been participating in western primate field studies since the 1950s, this shift coincided with a greater number of women becoming primatologists, and also some of them formulating their motivations in explicitly feminist terms.[217] Self-declared feminist primatologists, such as Jeanne Altmann, Sarah Blaffer Hrdy, Thelma Rowell, Linda Fedigan, Barbara Smuts and Meredith Small, explicitly declared that they choose to register sex differences, study females and use the "female point of view" as a feminist strategy in their work. Thereby they shifted the scientific priorities and affected others in the field to do the same—to study females as active agents in evolution.[218] Donna Haraway believes that "a unifying theme in the primatology done by women has been their high likelihood of being skeptical of generalizations and their strong preference for explanations full of specificity, diversity, complexity, and contextuality."[219] Women usually took the lead in changing assumptions and narratives about female primates, but a number of men did too.[220] Science historian and philosopher Sarah Richardson points out that independent of whether a researcher identifies as a feminist, their work may be identified as gender-reflective, that is "intellectual practice that works to "make gender visible", to "reveal gender", or to "prevent gender from being disappeared" in an area of intellectual inquiry."[221] Thus, a number of men primatologists also used gender-reflective approaches. Pioneering feminist primatologists were working actively with gender-reflective approaches, which became incorporated into mainstream primatology, even if these feminist insights are often not explicitly described as such, but seen as ways of making better science.[222] As Donna Haraway's analysis of primatology showed, destabilizing traditional stories becomes possible in a social context where different meanings are negotiated in other areas of life, such as women's role in society, because destabilization of narratives is a collective endeavor.[223]

---

[217] Fedigan, 2001: 55.
[218] Fedigan, 2001: 49.
[219] Haraway, 1989: 397.
[220] Haraway, 1989: 293ff; Fedigan, 2001: 59ff.
[221] Richardson, 2013: 126.
[222] Fedigan, 2001.
[223] Ibid.: 303.

Linda Fedigan suggests that North American primatology has been unusual, few other fields subjected to feminist critique have self-corrected so rapidly and to such an extent.[224] Likewise, Hrdy notes that at least in US primatology, there has been little resistance against acknowledging bias and the field has been responsive to criticism and self-correcting.[225] She states that rather than just a shift toward including females, there was a stampede to study them—"by the end of the 80s, female strategies were center stage".[226]

Moreover, Fedigan notes that North American primatology does not, in contrast to other field sciences such as paleontology, archeology and geology, have a macho field culture. Women were included early on and they showed that they were competent to conduct field as well as laboratory studies, laying the ground for a non-macho culture, where both women and men are included, in which feminist ideas could flourish.[227] Feminist science scholarship has called attention to androcentrism in male-dominated sciences, the coinciding revelation of androcentrism in science with the influx of women in primatology and other fields, and the role of feminism in the criticism of taken-for-granted background assumptions.[228] Thus, feminist analysts of science have demonstrated the role of the researchers' situatedness in terms of gender, race, socioeconomic status, sexuality and other social categories for their research questions, observations and hypotheses.[229]

In addition to primatology, science historian Londa Schiebinger and science philosopher Lynn Hankinson Nelson have explored several parallel examples of how feminist critique has contributed to the advancement of scientific knowledge in archaeology, developmental biology and medicine.[230] Londa Schiebinger notes that in both primatology and medicine the majority of feminist interventions have been on reevaluations of females, that is in my terminology, female turns. One recent analysis of

224 Fedigan, 2001: 48.
225 Hrdy, 1986: 141.
226 Hrdy, 1981/1999: xvii.
227 Fedigan, 2001: 55.
228 Nelson, 2017: 97.
229 Haraway, 1989, Harding, 1986.
230 Schiebinger, 1999, Nelson, 2017.

such a female turn in science is Sarah Richardson's investigation of the history of sex-determination science.[231] She traced the history of research looking for sex-determining genes in the 1980s. In line with the stereotypical binary assumptions that genes on the X chromosomes were female specialized and those on the Y chromosomes male specialized, researchers looked for genes involved in male reproduction in the Y chromosome. After identifying a sex-determining gene on the Y-chromosome in 1990, the *SRY* gene, contrary to expectations research showed that it did not function to control or activate a pathway, neither did they find a gene target for it in the testes, but it turned out to be closely related to genes on the X-chromosome. The prevalent androcentric "master gene" model of sex determination was not corroborated. Richardson shows that feminist critiques of these androcentric assumptions contributed to the development of a novel model of sex determination in which several genes are involved in sex determination and in that researchers started studying the genetics of the development of ovaries as well as testes.[232] Richardson describes the three phases of change, first cultural change in and around the field opened up for scientists' receptivity of gender-reflective approaches, then respected sex-determination geneticist Jennifer Graves started to employ a feminist framework and developed an alternative model, and lastly, the scientific community started to see gender-reflective approaches as productive and incorporated into mainstream sex-determination research practices.[233] Thus, feminist critique and gender-reflective approaches has contributed to progress in the content of science in sex-determination science as well as in primatology.

## Chapter Conclusions

As primates are our closest living relatives, the field studying them has rendered extensive public and media attention, attracted researchers from many disciplines and become a field ripe in controversies and

[231] Richardson, 2013.

[232] Ibid.: Chap. 7.

[233] Ibid.: 131.

negotiations about sex differences, human evolution, the origin of the heterosexual family and sexual division of labor.[234] Primates as models for human evolution have brought about different meanings in different cultures. Japanese primatologists tried to figure out how the whole "species society" and social relations functioned and realized already in the 1950s that dominance hierarchies in Japanese macaques were mostly dependent on strong kinship bonds among female relatives.[235] Japanese primatology does not seem to have started out with assumptions about passive females, which their western counterparts did to a large extent. Western primatologists focused on dominating and aggressive males and held theoretical assumptions of coy, passive, monogamous and caring females. Furthermore, they over-focused on the role of male hunting in human evolution and theorized female orgasm as a by-product of selection for male orgasm. In this history of a female turn in western primatology, I have found several processes of ignorance production: the repeated pattern of male precedence; lack of citation of the early studies that described females as sexually active and the importance of female kin relations for dominance; excluding evidence of non-human primate female orgasms from laboratories; androcentric reception of balanced empirical studies; reinterpreting suggestions of autonomous and innovative women into the prevalent androcentric story of women being dependent on men; and undermining the authority of and not citing Japanese primatologists—who early on recognized females as powerful in Japanese macaque societies. Most of these processes of ignorance production were due to the androcentric focus of the field and all delayed or distorted knowledge about active females. Many of these processes are no longer operating or much diminished, but in the wider field of sexual selection research, there are current examples of male precedence and continued production of ignorance about sexual selection in females (see Chap. 7).

Yet, already from the start, the community of primatologists has held heterogeneous views about females. Primatologists' perspectives have differed—by their cultural context, their feminist insights, their institutional position and their study species—and formed their situated

---

[234] Haraway, 1989.

[235] Ibid.: 253.

knowledges of females and sexual selection. In early western primatology, the domination of men in research is connected with the androcentrism in the field and hence a situated ignorance about females, although there are exceptions. The influx of women and gender-reflective approaches shifted research to include attention to females in the scientific community. Feminist primatologist Sarah Hrdy's situated knowledges were formed by her empathy with female langurs whose offspring got killed by males, feminist insights, and her institutional position at the heart of the developing field of sociobiology. Similarly, Meredith Small was engaged in female independence and agency, influenced by feminist insights and sociobiology when focusing on female choice and strategies among different species of macaques. Both Hrdy and Small challenged assumptions about passive and sexually reserved females. However, depending on their specific contexts, their negotiations of scientific knowledge and experiences of the reception of their results by the scientific society differ greatly. Sarah Hrdy, who pioneered feminist critique and interventions at a time when western primatology was male-dominated and androcentric, experienced controversy and resistance, as well as support. Meredith Small, who started her research later and right at one of the nodal points for feminist interventions in primatology, experienced no resistance related to her focus on females in her investigations. Small discuss the role of specific study species for her understanding, and this role is also illustrated by the alternative perceptions of primate society that developed from studying New World monkeys and lemurs. The perceptions of females as powerful and active in Japanese primatology, in contrast to western primatology, emphasize the role of cultural context for researchers' situated knowledges.

Japanese primatologists pioneering findings of powerful female primates were long disregarded by western primatologists, yet compared to other evolutionary biologists, western primatologists were early to explore female active sexual strategies. Feminist primatologists criticized the prevailing androcentric assumptions about coy, passive and monogamous females based on the accumulating evidence that female primates often initiate mating with several males. A number of pioneering feminist primatologists' interventions lead to methodological refinement, investigations of female lives and forwarding of new perceptions of females as active, aggressive, dominant, and pursuing their own strategies.[236] Over time, gender-reflective approaches came to be seen as useful in the scientific primatology community and was incorporated into mainstream primatology practice.

## References

Ah-King, M. (2022). The history of sexual selection research provides insights as to why females are still understudied. Nature Communications 13(1) 6976 https://doi.org/10.1038/s41467-022-34770-z

Ah-King, M., Barron, A., & Herberstein, M. E. (2014). Genital evolution: Why are females still understudied? *PLoS Biology, 12*(5), e1001851.

Alberts, S. C., & Silk, J. (2013). The contributions of Jeanne Altmann. *Evolutionary Anthropology, 22*, 198–199.

Altmann, J. (1974). Observational study of behavior: Sampling methods. *Behaviour, 49*, 227–267.

Asquith, P. (1996). Japanese science and Western hegemonies. In L. Nader (Ed.), *Naked science: Anthropological inquiry into boundaries, power and knowledge* (pp. 239–256). Routledge.

Asquith, P. (2000). Negotiating science: Internationalization and Japanese primatology. In S. C. Strum & L. M. Fedigan (Eds.), *Primate encounters: Models of science, gender and society* (pp. 165–183). University of Chicago Press.

Bagemihl, B. (1999). *Biological exuberance: Animal homosexuality and natural diversity*. Profile.

Bernstein, I. S. (1981). Dominance: The baby and the bathwater. *Behavioral and Brain Sciences, 4*, 419–457.

Clutton-Brock, T. H. (1978). Reproductive strategies in langurs. *Nature, 273*, 325.

Clutton-Brock, T. H. (1983). Behavioural ecology and the female. *Nature, 306*, 716.

de Waal, F. (2003). Silent invasion: Imanishi's primatology and cultural bias in science. *Animal Cognition, 6*, 293–299.

Dixon, A. F. (1977). Observations on the displays, menstrual cycles and sexual behaviour of the "Black ape" of Celebes (*Macaca nigra*). *Journal of Zoology, 182*, 63–84.

---

[236] Haraway, 1989.

Fedigan, L. M. (1982/1992). *Primate paradigms: Sex roles and social bonds*. University of Chicago Press.

Fedigan, L. M. (1983). Dominance and reproductive success in primates. *Yearbook of Physical Anthropology, 26*, 91–129.

Fedigan, L. M. (1984). Sex ratios and sex differences in primatology. *American Journal of Primatology, 7*, 305–308.

Fedigan, L. M. (1986). The changing role of women in models of human evolution. *Annual Review of Anthropology, 15*, 25–66.

Fedigan, L. M. (2000). Gender encounters. In S. C. Strum & L. M. Fedigan (Eds.), *Primate encounters* (pp. 499–520). University of Chicago Press.

Fedigan, L. M. (2001). The paradox of feminist primatology: The Goddess' discipline? In A. N. H. Creager, E. Lunbeck, & L. Schiebinger (Eds.), *Feminism in twentieth-century science, technology, and medicine*. University of Chicago Press.

Ganetz, H. (2004). Familiar beasts: Nature, culture and gender in wildlife films on television. *Nordicom Review, 26*(1–2), 197–214.

Haraway, D. J. (1986). Primatology is politics by other means. In R. Bleier (Ed.), *Feminist approaches to science*. Pergamon Press.

Haraway, D. J. (1988). Situated knowledges: The science question in feminism and the privilege of partial perspectives. *Feminist Studies, 14*(3), 575–599.

Haraway, D. J. (1989). *Primate visions: Gender, race and nature in the world of modern science*. Routledge.

Harding, S. (1986). *The science question in feminism: Industrial policy in Europe*. Cornell University Press.

Harding, S. (2006). Rethinking standpoint epistemology: What is 'strong objectivity'? In E. F. Keller & H. E. Longino (Eds.), *Feminism and science* (pp. 235–248). Oxford University Press.

Hrdy, S. B. (1974). Male-male competition and infanticide among the langurs (Presbytis entellus) of Abu, Rajasthan. *Folia Primatologica, 22*(1), 19–58.

Hrdy, S. B. (1977a). Infanticide as a primate reproductive strategy. *American Scientist, 65*(1), 40–49.

Hrdy, S. B. (1977b). *Langurs of Abu: Female and male strategies of reproduction*. Harvard University Press.

Hrdy, S. B. (1981/1999). *The woman that never evolved*. Harvard University Press.

Hrdy, S. B. (1986). Empathy, polyandry, and the myth of the coy female. In R. Bleier (Ed.), *Feminist approaches to science*. Pergamon Press. Reprinted in J. A. Kourany (Ed.), *The gender of science* (pp. 171–191). Pearson Education/

Prentice-Hall, 2002; reprinted in E. Sober (Ed.), *Conceptual issues in evolutionary biology* (pp. 131–159). Bradford Books of MIT Press, 2006.

Hrdy, S. B. (1988). Raising Darwin's consciousness: Females and evolutionary biology. In R. Bellig & G. Stevens (Eds.), *The evolution of sex* (pp. 161–169). Harper & Row Publishers.

Hrdy, S. B. (2000). *Mother Nature: A history of mothers, infants, and natural selection*. Pantheon.

Hrdy, S. B. (2009). *Mothers and others: The evolutionary origins of mutual understanding*. Belknap Press of Harvard University Press.

Hrdy, S. B. (2010). Myths, monkeys, and motherhood: A compromising life. In L. C. Drickamer & D. A. Dewsbury (Eds.), *Leaders in animal behavior, second generation* (pp. 343–374). Cambridge University Press.

Hrdy, S. B., & Hrdy, D. B. (1976). Hierarchical relations among female Hanuman Langurs (Primates: Colobinae, *Presbytis entellus*). *Science, 193*(4256), 913–915.

Hrdy, S. B., & Williams, G. C. (1983). Behavioral biology and the double standard. In S. K. Wasser (Ed.), *Social behavior of female vertebrates*. Academic Press.

Jay, P. (1963). The female primate. In S. Farber & R. Wilson (Eds.), *The potential of woman* (pp. 3–47). McGraw Hill.

Kawamura, S. (1958). The Matriarchal social order in the Minoo-B group: A study on the Rank System of Japanese Macaque. *Primates, 1*, 149–156.

Keller, E. F. (1983). *A feeling for the organism: The life and work of Barbara McClintock*. W. H. Freeman.

Linton, S. (1974). Woman the gatherer: Male bias in anthropology. In S.-E. Jacobs (Ed.), *Women in perspective: A guide for cross-cultural studies* (pp. 9–21). University of Illinois Press.

Lloyd, E. A. (1993). Pre-theoretical assumptions in evolutionary explanations of female sexuality. *Philosophical Studies, 69*, 139–153.

Nelson, L. H. (2017). *Biology and feminism: A philosophical introduction*. Cambridge University Press.

Parish, A. R. (1994). Sex and food control in the uncommon Chimpanzee—How Bonobo females overcome a phylogenetic legacy of male-dominance. *Ethology and Sociobiology, 15*(3), 157–179.

Parish, A. R., & De Waal, F. B. M. (2000). The other 'closest living relative'—How Bonobos (*Pan paniscus*) challenge traditional assumptions about females, dominance, intra- and intersexual interactions, and hominid evolution. In D. LeCroy & P. Moller (Eds.), *Evolutionary perspectives on human reproductive behavior* (pp. 97–113). Annals of the New York Academy of Sciences.

Rees, A. (2009). *The infanticide controversy: Primatology and the art of field science*. University of Chicago Press.

Richardson, S. S. (2013). *Sex itself: The search for male and female in the human genome*. University of Chicago Press.

San Diego Zoo. (2021). Vervet monkey (*Chlorocebus pygerythrus*) fact sheet: Behavior & ecology. https://ielc.libguides.com/sdzg/factsheets/vervet

Schiebinger, L. (1999). *Has feminism changed science?* Harvard University Press.

Segerstråle, U. C. O. (2000). *Defenders of the truth: The battle for science in the sociology debate and beyond*. Oxford University Press.

Small, M. F. (Ed.). (1984). *Female primates: Studies by women primatologists*. Alan R. Liss.

Small, M. F. (1989). Female choice in nonhuman primates. *Yearbook of Physical Anthropology, 32*, 103–127.

Small, M. F. (1990). Promiscuity in Barbary Macaques (*Macaca sylvanus*). *American Journal of Primatology, 20*, 267–282.

Small, M. F. (1993). *Female choices: Sexual behavior of female primates*. Cornell University Press.

Sterck, E. H. M., & Goossens, B. M. A. (2008). The meaning of "macaque" facial expressions. *Proceedings of the National Academy of Sciences U S A, 105*(43), E71. https://doi.org/10.1073/pnas.0806462105

Strum, S. C., & Fedigan, L. M. (2000). Changing views of primate society: A situated North American view. In S. C. Strum & L. M. Fedigan (Eds.), *Primate encounters: Models of science, gender and society* (pp. 3–49). University of Chicago Press.

Symons, D. (1979). *The evolution of human sexuality*. Oxford University Press.

Symons, D. (1982). Another woman that never existed: A review of the woman that never evolved. *The Quarterly Review of Biology, 57*(3), 297–300.

Takasaki, H. (2000). Traditions of the Kyoto School of Field Primatology in Japan. In S. C. Strum & L. M. Fedigan (Eds.), *Primate encounters: Models of science, gender and society* (pp. 151–164). University of Chicago Press.

Tuana, N. (2004). Coming to understand: Orgasm and the epistemology of ignorance. *Hypatia, 19*(1), 194–232.

Tuana, N. (2006). The speculum of ignorance: The Women's Health Movement and epistemologies of ignorance. *Hypatia, 21*(3), 1–19.

Van Schaik, C., & Silk, J. B. (2013). Contributions of Sarah Blaffer Hrdy. *Evolutionary Anthropology, 22*, 200–201.

Wrangham, R. W. (1980). An ecological model of female-bonded primate groups. *Behaviour, 75*, 262–300.

Wright, E., Grawunder, S., Ndayishimiye, E., Galbany, J., McFarlin, S. C., Stoinski, T. S., & Robbins, M. M. (2021). Chest beats as an honest signal of body size in male mountain gorillas (*Gorilla beringei beringei*). *Nature Scientific Reports, 11*, 6879.

Zuckerman, S. (1932/2015). *The social life of monkeys and apes*. Routledge.

# 3

# Turning Appreciation of Female Choice

As a background to the female turn in sexual selection research, this chapter concerns the turning appreciation of animal female choice through the duration of the controversy over animal female choice as a mechanism of sexual selection. The chapter includes a historical perspective on the changing views of sexual selection theory itself, and charts how perceptions about females have varied depending on the individual's situated knowledges. This controversy started with Darwin's contemporary scientists' skeptical reactions toward sexual selection theory and ended around 1980. Several science historians have contributed to our understanding of this history. In her definitive study *Darwin and the Making of Sexual Selection* (2017),[1] Evelleen Richards provides a meticulous contextual historical analysis of how Darwin developed his theory of sexual selection and how it was received in science and society. In *From Eve to Evolution: Darwin, Science, and Women's Rights in Gilded Age America* (2014),[2] Kimberly Hamlin gives a detailed account of how American feminists both engaged in and criticized sexual selection theory in the late nineteenth century. Furthermore, in *Looking for a Few Good*

[1] Richards, 2017.

[2] Hamlin, 2014.



M. Ah-King, *The Female Turn*, https://doi.org/10.1007/978-981-19-7161-7_3

*Males: Female Choice in Evolutionary Biology* (2010),[3] Erika Milam, in her intricate historical analysis focusing on struggles between disciplines, traces the partly diverging histories of female choice and sexual selection, from Darwin until the 1970s.

First, this chapter draws on Richards', Hamlin's and Milam's earlier research to write a historiography of females in sexual selection, the controversy over female choice in animals as a mechanism of sexual selection and the nineteenth-century feminist engagement in as well as critique of sexual selection. Secondly, I use my interviews with ornithologist and pioneer of female choice studies in birds, Malte Andersson, and rodent expert and prominent feminist critic of the prevailing Darwin-Bateman paradigm in sexual selection, Zuleyma Tang-Martínez, as well as the scientific record, to describe the ending of the controversy over female choice, the modern development of sexual selection theory and the critique thereof.

Throughout this chapter, I investigate how the researchers' situated knowledges[4] have formed their understanding of sex and sexual selection, as well as how ignorance about females has been produced. This history provides the setting for the female turn in sexual selection.

## A Contextual Gender History of Charles Darwin's Sexual Selection

In order to understand the female turn in evolutionary biology, it is important to comprehend the contextual gender history of Charles Darwin and his development of sexual selection theory. Science historian Evelleen Richards chronicles how the societal and political context in which Darwin formulated and presented his theory of sexual selection influenced his thinking.[5] Darwin worked within as well as against prevailing assumptions of sex and race in his time and place, namely England in the Victorian colonial era. With sexual selection theory, he tried to

[3] Milam, 2010.
[4] Haraway, 1988.
[5] Richards, 2017.

explain not only how differences between the sexes had evolved, but also how different human races had emerged. Darwin was a vehement abolutionist,[6] as described by historians Adrian Desmond and James Moore, and his hatred of slavery was an important motivation for his views of human evolution as of common descent, contrary to those who argued that human races were of different origins. Nevertheless, he described a racial hierarchy that has been used widely to justify racial conflict and ethnic cleansing.[7] Furthermore, his development of sexual selection theory occurred in the middle of the late nineteenth-century debates about "the woman question", that is campaigns for women's entry into universities, for married women's rights to own money and property (which was by law transferred to the husband at marriage), and for women's suffrage.[8] Conventional views held that men were intellectually superior to women, while women were morally superior to men, and that these sex differences were God-given.[9] Darwin had developed natural selection against the prevailing idea that God created differences among species. Likewise, he argued for a naturalistic explanation when investigating sex differences and developing the theory of sexual selection. These ideas were welcomed by feminist thinkers contesting explanations of male dominance in society as ordained by God.[10] Although Darwin admired British philosopher John Stuart Mill's political liberalism in *On Liberty*,[11] later when Mill, in *The Subjection of Women*,[12] extended liberalism to women and suggested that supposed moral and mental differences between men and women were due to women's lack of education and opportunities, Darwin discredited Mill with biological arguments.[13] In the *Descent of Man and Selection in Relation to Sex*,[14] Darwin argued in overt opposition to Mill that women were inherently

[6] Desmond & Moore, 2009: 1.

[7] Ibid: xix.

[8] Richards, 2017: 442ff.

[9] Ibid: 442.

[10] Hamlin, 2014.

[11] Mill, 1859.

[12] Mill, 1869.

[13] Richards, 2017: 446.

[14] Darwin 1871.

intellectually inferior to men.[15] Darwin claimed that sex differences in humans were the result of natural and sexual selection, that men's "struggle for possession of women" and greater participation in the struggle for survival meant that they had evolved to be larger, stronger and have higher intelligence and superior intellect.[16]

Richards persuasively argues that it was not mere prejudice that made Darwin couple feminine with primitive intelligence, instead, it was a logical effect of his theoretical arguments. In line with Haeckel's theory that individual development recapitulated species history, Darwin viewed women's more "childlike anatomy" as a manifestation that they belonged to an earlier developmental stage of the group, and he described women's abilities of intuition, rapid perception and nurturing as on par with the "lower" races—to Darwin, man was evolved woman.[17] Furthermore, Darwin's financially independent position meant that he had no salaried professional position to defend, so "he had no professed desire to keep women and the "lower" races in their place", but he viewed society's organization as a consequence of how nature works.[18] Thus, despite his vehement abolitionist stance, he sanctioned Thomas Henry Huxley (who called himself "Darwin's bulldog" because of his vigorous public support for the theory of natural selection and defense of it against criticism from the church), when Huxley argued that it was evolutionary impossible for women and blacks to reach the liberal ideal of equality.[19] Though, in contrast to Huxley, who worked openly against women's access to professional science, Darwin was not actively involved in sexual discrimination.[20] Thus, Richards asserts that, by Victorian standards, Darwin was a liberal man who encouraged a few women to make scientific observations and publish their work, and complimented the women at Cambridge when they finally gained the right to university examinations. However, his own personal experiences did not contradict the Victorian

[15] Richards, 2017: 441 f.

[16] Darwin 1871: part II: 326 f.

[17] Richards, 2017: 448.

[18] Ibid: 448.

[19] Ibid: 448.

[20] Ibid: 60.

perceptions of women as coy and nurturing,[21] nor his perception of indigenous people as inferior, rather, they corroborated his understanding of social inequalities.[22]

Since the beginning of his development of sexual selection theory, Darwin believed that the human races had evolved through the development of specific heritable standards of beauty, or aesthetic taste, among human groups.[23] He then extended this idea of aesthetic taste to animals and combined it with the conception of struggle for mates. Darwin's situated knowledges about sexual selection were formed by his experiences of traveling around the globe as a gentleman naturalist, his familiarity with the art of bird breeding, his own theories of inheritance and embryogenesis, and he worked both in line with and against contemporary Victorian ideals. Thus, his conviction of female choice in animals went against Victorian assumptions, although in line with those, he described female choice as rather passive and reluctant. Paradoxically, in accordance with Victorian tradition, Darwin assumed that male choice prevailed among humans.

As Richards convincingly demonstrates, disputes about sexual selection were inextricable from ideology and politics of race, sexuality and gender, and his theory provided potential solutions to current problems of the rulers of the British empire—the labor unrest, socialism, the suffragettes and the proliferation of the "unfit".[24] In Richards' words, "Darwin showed the way in the *Descent [of Man]* with his promotion of a racial hierarchy, his demonization of the savage, his explicit justification of feminine domesticity and intellectual inferiority, and hints toward racial improvement through better informed choice of marriage partners".[25] Thus the far-reaching societal legacy of the *Descent of Man,* even despite the controversy and scant support for sexual selection theory. In England, as the theory received little support from naturalists and biologists by the turn of the nineteenth century, it played a more vigorous

[21] Ibid 60–61.

[22] Ibid: 448.

[23] Ibid: 518.

[24] Ibid: 526.

[25] Ibid: 526.

role among novelists, utopians, feminists, political radicals and sex reformists, making it even more problematic for professional Darwinists.[26] Likewise, historian Kimberly Hamlin argues that it was partly because of the embracement of sexual selection—as an alternative to God-given male dominance in society—by freethinkers, socialists and feminists that the American naturalists were skeptic to the theory.[27] Among the public audience, Hamlin notes, "for nineteenth-century men and women, the phrase 'sexual selection' and the title *The Descent of Man* often functioned as shorthand for new ideas about gender roles and courtship".[28]

## The Darwin-Wallace Debate over Female Choice

A debate over sexual selection between Charles Darwin and his co-discoverer of natural selection, Alfred Russel Wallace, foreboded the difficulties the theory met in the scientific community later on. Evelleen Richards recounts that according to the Victorian norms of courtship and sexual practice, only males were believed to have sexual preferences. Therefore, Darwin's suggestion that, for one, female animals were mainly exerting sexual preferences and also that animals have a sense of beauty, was a twofold challenge to prevailing perceptions. Alfred Russel Wallace was a persistent critic of sexual selection and in 1868 (three years prior to the publication of *the Descent of Man*), he and Darwin had a long dispute mailing letters back and forth,[29] which ended with their mutual agreement to disagree—they could not convince each other.[30]

Darwin based his ideas of female choice on his understanding of inheritance and development—a male-centered view of highly developed males (in contrast to less developed females) transmitting their superiority in strength, intelligence, weapons and beauty mainly to their male

[26] Ibid: 492–493.

[27] Hamlin, 2021.

[28] Ibid: 125.

[29] Richards, 2017: 185 f.

[30] Ibid: 408.

offspring.[31] He considered males to be active in battling, displaying and mating, and females, because of their lesser "passion", to be passive, discriminating choosers.[32] To Darwin, because he relied heavily on what he had learned from the art of breeding birds—just as pigeon breeders chose by their aesthetic taste and modified features, so would females have chosen more and more ornamented male birds—it was male ornamentation that needed explanation, not female dullness in plumage[33] (in contrast to Wallace's perspective).

Wallace had long experience in collecting birds, butterflies and other specimens in the tropics in order to finance his research.[34] His research focus was on camouflage, how animals had evolved to blend in with their surroundings, and he thus explained the evolution of cryptically colored females through protection from predation, that is natural selection.[35] Furthermore, he compiled a list of bird species in which both males and females were conspicuously colored, and based on this list, he suggested that when females nest in protected places (like holes in trees), they are as beautifully colored as males. He concluded that in these cases, sexual selection worked on both sexes.[36] Thus, to Wallace, natural selection of females was responsible for the evolution of sex differences in plumage. The elaborate male traits he explained as an effect of surplus energy and physiological development. To him, the protective coloration of females was more important for the future of the species, the elaborate males he thought were in most cases unnecessary.[37] Although Wallace initially gave some credit to the idea of female choice in animals, over time he came to the conclusion that it was not necessary for explaining the observed patterns in nature.[38]

[31] Ibid: 186.
[32] Ibid 186.
[33] Ibid 186.
[34] Ibid: 407 f.
[35] Ibid: 393.
[36] Ibid: 393.
[37] Ibid: 402.
[38] Cronin, 1991: Chap. 6.

Darwin considered female choice in birds as an instinctive appreciation of beauty (thus an unconscious process[39]). Wallace, on the contrary, argued against female choice as a conscious process,[40] as did the majority of his contemporaries. Instead, they considered aesthetic taste a uniquely human ability and thereby rejected female choice in animals.[41] To Wallace, humans were unique in their ability to raise above the all-encompassing competition and develop sympathy, cooperation and altruism, a conviction which led him to correspond with John Stuart Mill, engage in land reform and women's rights.[42] His positive impression of indigenous people, their beauty, grace and mind, led him to view natural selection as not enough to explain human moral sense and intelligence, to him it must be due to an "overruling intelligence"[43]. Wallace's beliefs in human equality, social reform, and spiritualism developed into a radical egalitarianism which included feminist ideas, although with retained Victorian ideas about female sexuality: he was for marriage and against free love.[44] In 1890, Wallace after long having criticized sexual selection, suddenly argued for free female choice in humans, he had developed a socialist positive eugenics vision, in which he argued that the progress of the human race depended on reforming society (by land reform and equal rights) so that women could express their true choice.[45]

Thus, Wallace's situated knowledges (not recognizing sexual selection in animals) were formed by his experiences of collecting insects and butterflies in the tropics and studying their camouflage, to him, female animals were the ones who had evolved their dull plumages through natural selection, and there was no need to explain male extravagant traits through female choice. However, eventually, based on his socialist egalitarian perspective, he supported female choice in humans.

Thus, Darwin was persistent in his belief that female choice was the selection force behind the extravagant traits exhibited by many males,

[39] Richards, 2017: 408; 598n25.

[40] Ibid: 439.

[41] Ibid: 402.

[42] Ibid: 404.

[43] Ibid: 405.

[44] Ibid: 401.

[45] Ibid: 401–402.

despite the skepticism shown by his contemporary colleagues. Females as agents of selection were important to Darwin, although paradoxically, he reserved mate choice to males in humans.[46] To Darwin, male extravagant traits were derived, thus dull-colored females did not need explanation—to Wallace, it was the protective coloration of females that needed explanation, while male coloration was only due to surplus energy. In this complex production of knowledge and ignorance about females, Darwin emphasized females as agents of sexual selection producing male extravagant traits, while Wallace stressed natural selection acting on females resulting in protective coloration. This controversy was not only a private matter between Darwin and Wallace, but continued in the scientific community.

## The Scientific Community's Response to Sexual Selection

Following his letter-wise failure to convince Alfred Russel Wallace about the importance of sexual selection, Charles Darwin further received little support for his theory even among fervent supporters of natural selection. Evelleen Richards' historical analysis shows that soon after the publication of *the Descent of Man and Selection in Relation to Sex* in 1871, Darwin realized that most of his contemporary scientists did not agree with him about sexual selection, they did not believe in female choice because they thought female animals could not have an aesthetic sense.[47] In the Victorian society, the sense of beauty was thought to be uniquely human. Furthermore, few agreed with Darwin's idea that sexual selection through diverging tastes of beauty could explain human racial diversity.[48] Alfred Russel Wallace and St. George Mivart were the leading scientific critics of the theory. Wallace in his review of *the Descent of Man*, approved of male-male competition, and accepted female choice to a degree in female birds, that they could appreciate ornaments and reject and choose

[46] Ibid: xxi.
[47] Richards, 2017: 466; Milam, 2010: 3.
[48] Richards, 2017: 467.

certain males, but he did not accept it among insects.[49] Wallace's priority was to assert the dominant role of natural selection. St. George Mivart, a former Darwinian who had turned into a Roman Catholic, had earlier written a row of papers criticizing natural selection and arguing for higher guidance of evolution, now charged Darwin for undermining social stability with his tale of animal origins of mind and morality.[50] Mivart too accepted male-male competition, but as a kind of natural selection, while rejecting female choice outright. As a primate expert, he used primates as examples to argue that the larger and stronger males did not leave any room for female choice.[51] In response, "Darwin's bulldog" Thomas Henry Huxley set out to defend natural selection, but said nothing in defense of sexual selection.[52]

Thus, the theory of sexual selection received little support among Darwin's contemporary scientists, even among his natural selection supporters. However, Darwin himself was not shaken in his conviction that sexual selection was the decisive explanation for the evolution of male extraordinary ornaments.[53]

## Early Feminist Appreciation and Critique of Sexual Selection

Outside of the scientific community, some people received sexual selection theory with enthusiasm. Charles Darwin, as the Victorian gentleman he was, had, together with his wife Emma and their daughter Henrietta,[54] scrupulously scanned the formulations in *the Descent of Man* to avoid indecency, yet his subject—sex and sexuality—lead to a number of different readings, and eugenicists, social purists and feminists endorsed female choice in their own ways.[55]

---

[49] Ibid: 468.
[50] Ibid: 469.
[51] Ibid: 470.
[52] Ibid: 472.
[53] Ibid: 478.
[54] Ibid: 440–441.
[55] Ibid: 482 f.

God-given male superiority was the prevailing explanation for women's role in western society during the nineteenth century. Historian Kimberly Hamlin in her thorough analysis of the connections between feminism and evolutionary science in Gilded age America (1870 until the turn of the century) shows that the introduction of Darwin's evolutionary theory inspired new thinking about gender and sex, and gave hope to feminists championing for women's rights.

Darwinian theory overturned Eve's curse as the predominant explanation for women's subjugation, and although scientific arguments and investigations to support male superiority followed suit,[56] evolutionary theory was multivalent. Women made their own interpretations of what a continuum between animals and humans meant, using the variety of "domestic arrangements" among animals as an argument for other possibilities for human society. Hamlin shows that the rapid societal changes at this time—after the Civil War there was a lack of male workers for the expanding industries, fewer men that could go to college and fewer men to marry—opened up opportunities for women to work, get an education and pursue activities outside home.[57] This development led to an increment of professional, usually single, women.[58]

The interest in science education among women grew. In 1886, Smith College in Northampton, Massachusetts built the first laboratory facilities to educate women in the sciences, and there was a surge of interest in reading and discussing scientific ideas in women's club networks as well as writing in popular scientific journals.[59] In these journals, prominent scientists and amateurs published side by side and debated new ideas. Hamlin's research showcase women's enthusiasm for evolutionary science, that they were engaged in science and considered it an ally, which resulted in debates between male scientists and pro-science feminists. While these feminists used science to question the status quo, in contrast, male scientists referred to science to defend the status quo of gender

[56] Russett, 1989.

[57] Hamlin, 2014:70.

[58] Ibid: 123.

[59] Ibid: 63ff.

relations, assuming them to be natural.[60] In a small window of time (1870s–80s), before the scientific establishment became professionalized and excluded women, some women were actively contributing to science and scientific debates and assisted in making objective scientific methods the norm.[61]

American pioneering Protestant minister and women's rights activist Antoinette Brown Blackwell, in *The Sexes Throughout Nature* (1875), published just four years after the publication of *Descent of Man,* was the first woman to criticize sexual selection theory. She argued that Darwin and his male colleagues ignored the special characteristics that females had evolved, and in focusing on male characteristics used the theory of male superiority as a "forgone conclusion".[62]

She pointed out the androcentrism[63] (the practice of placing the male point of view in the center of one's worldview) in research as an effect of the science being produced by male researchers. "Standing on a learned masculine eminence, looking from their isolated male standpoints through their men's spectacles and through the misty atmosphere of entailed hereditary glamour" they could only see evidence of women's "natural" inferiority.[64] Thus, she provided a gender-reflective approach, making visible how gender conceptions operate in knowledge.[65] Blackwell was well aware that she did not have the same authority as those she criticized:

> "I do not underrate the charge of presumption which must attach to any woman who will attempt to convert the great masters of science and of scientific inference. But there is no alternative! Only a woman can approach the subject from a feminine standpoint; and there are none but beginners among us in this class of investigations".[66] Blackwell argued for objective

[60] Ibid: 69.

[61] Ibid: 169.

[62] Blackwell, 1875:14.

[63] Androcentrism is the practice of placing the male point of view in the center of one's worldview.

[64] Blackwell, 1876: 1.

[65] Richardson, 2013:17.

[66] Blackwell, 1875: 22.

scientific methodology to perform accurate studies of women, in which women's experiences should be used to form a new science to understand the differences between women and men.

At a time when the defining lines of science were under negotiation, moving away from including amateurs and toward distinguishing professional science, Darwinian feminists were engaged in what science sociologist Thomas Gieryn calls boundary work in negotiating the borders between science and non-science.[67] They criticized biological arguments for female inferiority based on female physiology, brain size and evolutionary arguments of women being less developed than men, and demanded objective methods and interpretations. Hamlin shows that women were actively contributing to creating new meanings of science, although the emerging scientific establishment proceeded to exclude them.[68]

When well-respected eye and ear physician and Harvard professor Edward Clarke in his influential *Sex and Education, or a fair chance for the girls* (1873)[69] claimed that women's reproductive functions were threatened by intellectual work, and thereby justifying the exclusion of women from education and science, it was met with vehement resistance by feminists.[70] They questioned his methods, demanded evidence, required female physicians and the right to speak for themselves, conducted studies and used their own experiences as evidence that intellectual and professional work did not hinder reproduction.[71] Questioning Clarke's dependence on anecdotes, they called for more scientific studies. Furthermore, Blackwell and others countered with describing her own good health while pursuing studies.[72] The debate did not only concern women's possibilities to study, but also their ambitions to speak in public and work outside their homes. Using varying animal examples of domestic and sexual arrangements and her own experiences, Blackwell argued

[67] Gieryn, 1999: 4–5.
[68] Hamlin, 2014: 67.
[69] Clarke, 1873.
[70] Ibid: 73ff.
[71] Ibid: 75ff.
[72] Ibid: 76.

for husbands to take part in household duties in order to enable women to combine family life with work.[73]

In the United States at this time, the Darwinian feminists consisted of educated, white women that—although they varied in their ideas about race—often adopted conventional ideas about racial hierarchy and argued that white educated women should be given the same rights as white educated men. Freethinking feminist Helen Hamilton Gardener joined the science classes of 1870s, and became a public debater in the brain size debates of 1880s and onwards arguing together with other prominent feminists—against neurologist William Hammond's assertions that women's brains were inferior to men's in nineteen ways—that female brain physiology did not limit women's mental abilities.[74] For example, she challenged Hammond to distinguish the sex of twenty brains by sight, but he did not comply. Gardener also questioned Hammond's results on the basis of the selection of the brains he compared—the brains of anonymous women who had died in hospitals, compared to renowned educated men's. At her death she donated her own brain to science, to support the investigation of able-educated women, in line with evolutionary ideas of racial hierarchy and her aim to elevate educated white women to the same opportunities and rights as similar men. She also pointed to interactions between nature and culture—and the possibility that opportunity and environment affected brains—thereby challenging biological determinism.[75]

In Europe, sexual selection engendered similar feminist responses. The British novelist Mary Ann Evans also known as George Eliot, provided an early feminist critique of evolutionary ideas as did Darwin's French translator, Clémence Royer, who in 1874 criticized male-exclusive science for assuming women to be passive, lacking instincts, passions and interests—in a paper that was withheld from publication by the journal editors.[76] Thus, the journal editors produced ignorance by censoring criticism. Royer popularized Darwinism in France through a number of

[73] Ibid: 101.

[74] Ibid: 78 f.

[75] Ibid: 91.

[76] Hrdy, 2000: 21–22.

evolutionary publications—and reinterpreted that which did not fit her own feminist and elitist views.[77] Swedish feminist Ellen Key was also an engaged Darwinian who argued (in contrast to American feminist Charlotte Perkins Gilman) that women should devote themselves entirely to maternity.[78]

Sexual selection provided women with scientific vocabulary and authority when arguing for female reproductive autonomy. Little did it matter that sexual selection by female choice was not generally accepted among scientists—Darwinian feminists argued that male mate choice in humans was unnatural—since it contrasted the principal pattern of female choice among animals. Therefore, enabling women's choice and autonomy would support evolutionary progress.[79] With the *Descent of Man*, Darwin had unwittingly put evolutionary theory right at the heart of popular and scientific debates about gender and sex, plus provided a scientific rationale for female reproductive autonomy.[80] Darwinian revolution even set the stage for the birth control movement in the United States,[81] as well as was used in the campaigns for birth control in the UK (which Darwin was against, he believed that there needed to be severe competition for existence in order for humans to advance).[82] However, the particular social and historical circumstances—the professionalization and masculinization of science—meant that these Darwinian feminists' criticism of male bias in the context of evolutionary science was ignored in the scientific community.

In response to women's organization for equal rights in the nineteenth century, white Victorian male scientists immersed themselves in research to determine scientific differences between men and women—a "sexual science" which did not judge in women's favor.[83]

[77] Richards, 2017: 398.

[78] Hamlin, 2014: 124.

[79] Ibid: 144.

[80] Ibid: 164.

[81] Ibid: 149ff.

[82] Richards, 2017: 494.

[83] Russett, 1989.

## The Eclipse of Sexual Selection

With the scarce support for sexual selection in the scientific community, the theory went into what has been described as an eclipse. Actually, after Charles Darwin's death in 1882, Darwinism in its entirety went into decline, not only was sexual selection out of favor, natural selection became highly controversial and reemerged only well into the twentieth century.[84] Science historian Erika Milam traces the partly different histories of female choice and sexual selection research in evolutionary biology, from Darwin until the 1970s.[85] Milam's analysis pays close attention to how boundaries between different groups of scientists, their lack of communication as well as competition between subdisciplines have contributed to disregard of the temporal stretched-out histories of female choice and sexual selection research. Here, I focus on how ignorance about females has been produced and maintained in the history of sexual selection.

### Marginalized Sexual Selection

In early twentieth century, natural selection was revived, but also controversy around sexual selection in close association with eugenics and questions about women's rights.[86] Research and promotion of sexual selection was continued by merely a few researchers, as the majority dismissed sexual selection as too contentious.[87] Male-male competition was subsumed under natural selection and the most controversial part, about female choice and animal aesthetic sense, was abandoned.[88] According to Evelleen Richards, "it was its connotations of greater freedom, independence, and agency for women in the context of the rise of the women's movement and emergence of the New Woman of the late nineteenth century that represented the greater challenge to the acceptance of sexual

[84] Richards, 2017: 524.

[85] Milam, 2010.

[86] Richards, 2017: 529.

[87] Ibid: 525.

[88] Ibid: 525.

selection".[89] Between 1900 and 1930, most professional zoologists rejected the idea of mate choice in animals, while restricting true mate choice to humans.[90] By 1900, not even the evolutionary biologists who supported natural selection believed in sexual selection. American zoologist and geneticist Thomas Hunt Morgan in 1903 heavily criticized and dismissed sexual selection. In his words, "the theory meets with fatal objections at every turn".[91] However, those specifically interested in human mate choice, namely early twentieth-century eugenicists and social hygienists, embraced sexual selection by female choice and argued that it was a means to improve humanity.[92]

Furthermore, like Alfred Russel Wallace, many nineteenth- and early twentieth-century scientists dismissed female choice in animals because they did not believe animals were able to make such rational decisions.[93] The prevailing explanation for female-male sexual interactions was that female animals were stimulated by courtship, and given enough stimulation would mate with a male.[94] When Stanford zoologist Vernon Kellogg in *Darwinism Today* (1907) reviewed the status of sexual selection, criticism of female choice dominated and he explained evolution of male armaments through natural selection. Kellogg failed to mention existing empirical support for female choice in animals, perhaps because it was presented by amateur naturalists rather than professional scientists.[95] Again, just as with the feminist Darwinians, the border between professionals and amateurs contributed to sustain ignorance about animal female agency.

For example, biology teachers and science enthusiasts George and Elisabeth Peckham studied mating behavior in spiders (1889–1890), testing Darwin's and Wallace's alternative explanations, and concluded that Wallace's argument of vigor as an explanation of elaborate male ornaments did not hold up. Although males were more ornamented,

[89] Ibid: 527.
[90] Milam, 2010: 30.
[91] Morgan, 1903: 221.
[92] Milam, 2010: 48.
[93] Ibid: Chap. 1.
[94] Ibid: 34.
[95] Ibid: 35.

females were more active. Furthermore, the most active jumping spiders were drab, and the most ornamented males belonged to sit-and-wait orb-web spiders (family *Araneidae*) who exposed their ornaments in the presence of females.[96] Thus, their research supported female choice of male ornaments and sexual selection.

Indeed, accounts of early pioneers in sexual selection testify how marginalized these ideas were—British ornithologist Edmund Selous in the early 1900s was disappointed by the lack of recognition for his pioneering contribution of recognizing active female choice in Ruffs, *Calidris pugnax*, and Black Grouse, *Lyrurus tetrix*: "I did everything in my power, to further scientific truth, and indeed have produced immensely strong evidence in favour of the Darwinian theory of sexual selection. It would seem, however, that, since the theory is (officially) out of favour, such evidence is not wanted."[97] Another example is Latvian ornithologist Hilda Cinat-Thomson, who in 1926 published a bird study showing that manipulation of the black throat spots of male budgerigars influenced their attractiveness to females—males with more spots were found to be more attractive—but as it was published in an obscure German journal, the study did not receive much attention.[98] Thus, notions of active females and studies corroborating sexual selection theory remained ignored.

In the 1930s and '40s, the "modern evolutionary synthesis" was developed which integrated the formerly incompatible frameworks of evolutionary theory, Mendelian genetics and population genetics—it did not include sexual selection. Thus, sexual selection was largely ignored by the scientific community at this time.

In the 1930s, Evelleen Richards describes how divergent social and political agendas lead British biologists Ronald A. Fisher and Julian Huxley (who were both involved in developing the evolutionary synthesis) to interpret sexual selection differently.[99] Fisher, a mathematician and geneticist at the agricultural research institution Rothamsted Experimental

---

[96] Ibid: 36.

[97] Cited in Birkhead et al., 2014: 329.

[98] Birkhead et al., 2014: 332.

[99] Richards, 2017: 529.

Station, developed his "runaway" sexual selection hypothesis, suggesting that a female preference for a trait can over time co-evolve with the preferred male trait (through a genetic correlation between preference and trait) and lead to exaggerated traits that "run-away" from control by natural selection. Fisher, with his eugenic perspective, emphasized choice of good mates as a means to avoid runaway effects; he meant that women should, through beneficial eugenic choice, marry courageous soldiers rather than be distracted by displays of wealth that could lead to middle-class degeneration. In the same line, Fisher's "sexual preference" also supported the maintenance of conventional gender roles, men should be chosen for bread-winner capacity and women for their abilities of housekeeping and raising children.[100]

In contrast, inspired by ideals of sexual equality, Oxford-educated zoologist Julian Huxley studied the courtship displays of monogamous crested grebes, *Podiceps cristatus* (birds), which show little morphological differences between the sexes. Initially, he reformulated sexual selection as "mutual sexual selection", which he meant led to transference of characteristics between the sexes and sex similarity.[101] (It is easy to see the influence of the crested grebes here.) Later on, in 1938, Huxley had changed his views and claimed that the majority of display characteristics could not be explained by sexual selection, but was instead related to species recognition and the excitement of female response to courtship, thus female choice was once again interpreted as natural selection.[102] In addition, the stimulatory notion of female choice again recast females in a passive role.

Fisher's work on sexual selection was largely overlooked at the time and only became well known after the republication of the *Genetical Theory of Natural Selection* in 1958.[103] In contrast, Huxley's subsuming of sexual selection under natural selection was influential and "cast a damper"[104] on sexual selection, which remained more or less overlooked until its revival in the 1970s.[105]

---

[100] Ibid: 530.

[101] Ibid: 530.

[102] Ibid: 530.

[103] Milam, 2010: 49.

[104] Geoff Parker interview 2017.

[105] Richards, 2017: 531.

## Bateman's Experiment

One of the few sexual selection experiments conducted during the eclipse was British geneticist and botanist Angus Bateman's, 1948 experiment on fruit flies, *Drosophila melanogaster*. In a for its time innovative experiment, Bateman tested Darwin's sexual selection predictions in experiments on fruit flies with different genetic markers, in order to trace the offspring of both parents. By letting a number of fruit flies procreate in vials for some time and later counting the resulting offspring, Bateman concluded that males vary more among themselves in number of mates and reproductive success than females, that females are choosy and males indiscriminate in mate choice—therefore he claimed, males are subject to stronger sexual selection than females.[106] He argued that "the specialization into large immobile gametes and small mobile gametes produced in great excess … would explain why … there is nearly always a combination of an undiscriminating eagerness in the males and a discriminating passivity in the females".[107] Bateman's study did not render much attention at the time of its publication, but later became a cornerstone of modern sexual selection. I will return to the criticism of this experiment and Bateman's conclusions later in this chapter.

Another account of the lack of interest in female choice as sexual selection comes from renowned evolutionary theorist John Maynard Smith: "When, in 1956, I published a paper showing, at least to my own satisfaction, how female fruit flies choose males, I do not remember receiving a single reprint request. My ethological contemporaries who were studying under Tinbergen[108] at Oxford refused to accept my explanation that some males failed to mate because they could not keep up with females during the courtship dance".[109] Among ethologists (researchers of animal behavior), the stimulation model of mate choice dominated,[110] females were thought to accept to mate as soon as they got enough species-specific stimulation by

---

[106] Bateman, 1948: 367.

[107] Ibid: 365.

[108] Nikolaas Tinbergen was one of the founders of the field of ethology, the study of animal behavior.

[109] John Maynard Smith in Cronin, 1999: ix.

[110] Milam, 2010: 55.

the male. Hence, female choice as a mechanism for sexual selection was a neglected topic among researchers in animal behavior in 1956, contributing to the ignorance about active females as well as sexual selection.

## Disciplinary Boundaries Affect Knowledge Transfer

The development in one discipline may affect the development in another field. For example, Erika Milam appoints population biologists' move toward studying mate choice within species—that is moving away from viewing mate choice as a mechanism for species recognition into perceiving it as a means to maintain genetic diversity within a population—as important for the further development of research in sexual selection.[111] However, due to the lack of communication between the disciplines, this population biology research was unknown among other behavioral researchers. According to John Maynard Smith, modern bird researchers studying behavior were not aware of the behavioral research on fruit flies.[112] Among population geneticists in the 1960s (as well as among biologists during the modern synthesis) female choice was mainly considered as a mechanism of species recognition.[113] Thus, Milam shows that biologists studying whole organisms (rather than molecules or cells) in the 1970s rarely cited the work of the fruit fly population geneticists, instead they refer to the evolutionary mathematical models of John Maynard Smith and Peter O'Donald.[114] The lack of communication between researchers of different fields is one way in which ignorance about females is maintained. On the other hand, connections to other research fields, such as Maynard Smith's and O'Donald's theoretical models connecting population genetics with animal behavior, can be one way in which knowledge about females transfers between disciplines. As we shall also see in Chaps. 5 and 6, interdisciplinary work has been one route toward gaining knowledge about active females and female multiple mating.

[111] Ibid: Chap. 5.

[112] Ibid: 130. John Maynard Smith recounted this in an interview to Erika Milam in 2002.

[113] Smith, 1991: 146.

[114] Milam, 2010: 131.

## Different Versions of Eclipse Narratives

Sexual selection theory, and female choice as a mechanism thereof, was largely marginalized for a long time. But biologists' interest in female choice as a mechanism for speciation (how new species emerge) and maintenance of genetic diversity was strong throughout the twentieth century.[115] However, there are several diverging accounts of when sexual selection reemerged: to quantitative theoretical biologists the eclipse ended when they worked on models of female choice for evolutionary change in the 1960s, and to population geneticists it ended when they studied female choice and the rare male effect (when females have frequency-dependent preferences and choose the male form that is rare) in the mid-1950s.[116] Erika Milam criticizes evolutionary biologists' "standard eclipse narrative", which states that the rebirth of sexual selection happened in the 1970s, for being a selective history.[117] She argues that this narrative served evolutionary biologists in their disciplinary struggles for prestige and funding in competition with the quickly expanding field of molecular biology—resulting in the exclusion of theoretical and laboratory biologists' contributions in the accounts of history of sexual selection.[118] Thus, Milam showed that interest in sexual selection was more stretched out in time and over research fields than what has commonly been believed.

# 1970s' Forceful Revival of Sexual Selection

Although the claim that sexual selection was "the only major scientific theory ever to have been accepted after a century of condemnation" is to exaggerate its disappearance,[119] during its history this line of research has often been marginalized and ignored by scientists. Erika Milam, as well as Evelleen Richards, persuasively contend that it was not until the 1970s

[115] Ibid: 160.

[116] Ibid: Chap. 5.

[117] Ibid: 6.

[118] Ibid: Chap. 6.

[119] Milam, 2010: 157–158.

that sexual selection theory forcefully came back in biological research, in the social and political context after the 1960s sexual revolution, psychological studies of William Masters and Virginia Johnson showing women's capacity for multiple orgasms, the abortion debates, the "second wave" feminist movement, the availability of the pill, and the entrance of many women in science.[120] Feminist critics, especially the women who had entered the male-dominated sciences were active in reformulating assumptions about females and putting sexually active females at center stage, as described in Chap. 2. In addition to this social and political context, other factors, such as development within the scientific community were also important for the increased interest in sexual selection theory. The "inside science" factors include the theoretical shift from group to individual selection, the rapidly successful Parental Investment theory, and the presentation of sociobiology. Although, as Donna Haraway keeps reminding, there are good reasons to question the distinction between factors "inside" and "outside" science influencing the development in a field.[121]

## Shifting from Group to Individual Selection

One of the factors that enabled a strong revival of sexual selection was the shift away from group selection (the notion that selection acts on whole groups of organisms, for example as selection "for the good of the species") and toward recognizing selection on the level of the individual, that is that selection acts on individuals and lead to individual adaptation. In the 1960s, most biologists assumed that traits evolved for the good of the species.[122] A book published in 1966 by American evolutionary biologist George C. Williams, *Adaptation and Natural Selection: A Critique of Some Current Evolutionary Thought,* in which he criticized the prevalent group selection thinking in evolutionary biology, marked a turning point toward a greater focus on individual selection. Williams claimed that "adaptation, when recognized, should be attributed to no higher level of

[120] Milam, 2010: 145f; Richards, 2017: 532.

[121] Haraway, 1989: 124 f.

[122] Simmons, 2001: ix.

organization than is demanded by the evidence",[123] thus rather invoking selection on individuals or genes than on groups.[124] He also provided an argument for why males would be more ardent than females, claiming that because females make greater physiological sacrifices in pregnancy, they should be coy and discriminating in comparison to males.[125] Other influential works for this shift were British ornithologist David Lack's *Ecological Adaptations for Breeding in Birds* (1968),[126] British entomologist[127] and theoretical biologist Geoff Parker's work on sperm competition[128] (see Chap. 6) and American evolutionary biologist Robert Trivers' (1972) influential chapter on Parental Investment,[129] described below.

## Parental Investment Theory

Thus, one important contributing factor for the increasing interest in sexual selection was Robert Trivers' Parental Investment theory which demands a separate presentation. He published his ideas in the chapter "Parental Investment and Sexual Selection" in the *Descent of Man* centenary celebration book *Sexual Selection and the Descent of Man, 1871–1971*.[130] It synthesized lessons from population geneticists, theoretical biology models and Angus Bateman's at that time overlooked experiment, and formulated testable predictions for field researchers to apply in their research.[131] Trivers argued that sex differences in choosiness was an effect of sex differences in parental investment—the sex that invested more in parental care would become scarce for the other sex which would then compete for mates.[132] Because females in most species provide more parental investment than males, this explained why females

123 Williams, 1966: v.

124 This book was later popularized by British evolutionary biologist Richard Dawkins in *The Selfish Gene* (1976).

125 Gowaty, 2003: 905.

126 Lack, 1968.

127 Entomologists are scientists studying insects.

128 Parker, 1970.

129 Trivers, 1972 described below.

130 Campbell, 1972.

131 Milam, 2010: 141.

132 Trivers, 1972.

would be discriminate and "coy". Trivers' model made it possible for field researchers to go out and test the predictions, thereby legitimizing female choice in the sexual selection research agenda and also made Angus Bateman's experiment reknowned. Furthermore, it motivated studies of "extra-pair paternity" in socially monogamous animals.[133] Herpetologist Mike Ryan recalls that already five years after its publication, Trivers' chapter was considered dogma in the scientific community.[134]

## Sociobiology

Lastly, the presentation of sociobiology and the following heated controversy contributed to increasing interest in sexual selection. It was entomologist and Harvard professor Edward O. Wilson who in *Sociobiology: the new synthesis* (1975) suggested a new theoretical framework to study social behavior from an evolutionary perspective. When applied to animals, this suggestion was uncontroversial, but in his last chapter Wilson hypothesized how social behavior in humans evolved and he described how "men forage for game or its symbolic equivalent in the form of barter and money",[135] while the woman chooses the man with the best genes and takes care of children. When sociobiologists claimed a biological basis for ardent sexually indiscriminate males and sexually passive females, the highly visible feminist movement vigorously contested these claims. Critical reactions to the sociobiologists' claims from feminists and other critics, notably Wilson's fellow Harvard professors Stephen Jay Gould and Richard Lewontin, and social scientists (who evidently had other explanatory frameworks for human social behavior), generated a long-standing scientific controversy which also produced massive media and popular interest in sexual selection.[136]

When interest in sexual selection increased in the 1970s, the question of whether female choice could act as a mechanism of sexual selection reemerged, ornithologist Malte Andersson recalls the frenzy around female choice at that time.

[133] Gowaty, 2003: 906.

[134] Mike Ryan interview 2017.

[135] Wilson, 1975: 553.

[136] Richards, 2017: 532. For a sociological analysis of the sociobiology controversy see Segerstråle, 2000.

## Malte Andersson—Pioneering Female Choice Studies in Birds

When I met him in his office, ornithologist Malte Andersson, Professor Emeritus in Animal Ecology at Gothenburg University, Sweden, described what led him to conduct his classical study on female choice in long-tailed widow birds, *Euplectes progne* (1982),[137] one of the first studies to show experimental field evidence of female choice. This experiment is often used in biology textbooks to illustrate female mate choice.

Andersson says that he had an education in engineering and mathematics and was since long an avid bird-watcher who preferred to be out in the field. Thus, rather than continuing with indoor work on the spread of radioactivity at the start of nuclear power use in Sweden, he chose to shift to zoology. As it was in the early days of the zoology department at Gothenburg University, his supervisor, an endocrinologist and a spare-time bird-watcher, was open-minded, and Andersson was allowed to choose his dissertation project freely. He studied breeding ecology, social behavior and communication in four different species of skuas, a group of predatory seabirds, inspired by his readings by evolutionary biologist George C. Williams and especially Dutch ornithologist Nikolaas (Niko) Tinbergen. Tinbergen was one of the grounding fathers of ethology (the study of animal behavior) who suggested that biological study of behavior should be investigated using four kinds of complementary questions concerning adaptive function, evolution, mechanism and individual development. Hence, Andersson's perspective was evolutionary and ecological explanations for behavior, focusing on questions such as: why do skuas have only two young per clutch? Are they limited by the production of eggs, by brooding, or by feeding of young? What is the relationship between prey abundance and breeding behavior?

After finishing his dissertation with the humble title *Some aspects of behaviour and ecology in skuas (Stercorariinae)* (1975),[138] he started studying sexual selection. The skuas had triggered his interest—he was contemplating on the skuas' long tails. One suggested hypothesis was species

137 Andersson, 1982.

138 Andersson, 1975.

recognition, but given that even an amateur ornithologist can distinguish the different species easily, he thought this explanation was not adequate. Besides, although male and female skuas look alike, when a pair is about to lay eggs, the male stands in front of the female and displays by lifting his tail up in the air—which suggested to Andersson that they had a sexual function (Fig. 3.1).

**Fig. 3.1** Malte Andersson in his office at the Zoological department, Gothenburg University, Sweden

## Ending the Female Choice Controversy

The combination of the development of theoretical models and empirical experiments led to the end of the controversy over female choice as a mechanism for sexual selection. Andersson says he worked on the most obvious problem in the end of the 1970s, when female choice was controversial. However, by that time there were no rabid opponents to female choice, recalls Andersson, although some researchers were skeptical.[139] For example, in one study on birds of paradise—currently considered a prime example of sexually selected plumages—the investigator was skeptical toward female choice having anything to do with the evolution of their plumage.

For others, female choice rendered interest and by the late 1970s, sexual selection was a hot topic.[140] In 1967, British ornithologist and theoretical biologist Peter O'Donald, published a population genetic model of mate choice, concluding that female choice could not lead to the evolution of elaborate traits that reduced survival. Andersson recalls that because O'Donald had claimed female choice for traits that were detrimental for survival was totally impossible,[141] and similarly, prominent theoretical biologist John Maynard Smith had argued that it was unlikely, a number of people were incited to simultaneously trying to prove them wrong. Andersson worked with a mathematical model of indicators, or what evolutionary biologist Amotz Zahavi called the "handicap principle",[142] that is how evolution of reliable signals can occur.[143] Through his work with this mathematical model, Andersson realized that a preference gene could increase in frequency for a trait even if it was disadvantageous to survival, contrary to O'Donald's claim. Likewise, the subsequent population genetic models by theoretical population

---

[139] Malte Andersson interview 2017.

[140] Milam, 2010: 157.

[141] O'Donald, 1967.

[142] Zahavi, 1975.

[143] If the expression of an extravagant trait comes with a cost, so that only individuals that good quality can afford to exhibit the trait (such as a peacock's tail)—then it becomes a reliable signal of quality.

geneticists Russell Lande (1981)[144] and Mark Kirkpatrick (1982)[145] showed that Ronald Fisher's runaway selection could work, that is a genetic correlation between female preferences and male sexual characteristics could lead to the evolution of extravagant male traits even if those traits are disadvantageous for survival.

The next step for Andersson was to test the female choice hypothesis experimentally. In 1975, after his dissertation, Malte Andersson went for a few weeks on vacation to Kenya where he saw long-tailed widowbirds—with the males flaunting their bizarre lengthy tails and the females looking like unconspicuous house sparrows—as well as the closely related Jackson's widowbirds *(Euplectes jacksoni)*, in which the ornamented males jump up and down in elaborate displays. After having dismissed peacocks as a study species (there were plenty of them in Sri Lankan national parks, but in rather dense vegetation and with dangerous buffalos and elephants around, the field conditions were too challenging), he returned to Kenya to try out experiments, cutting and pasting tail feathers of the long-tailed widowbirds. A successful test season was followed by another season and resulted in the article "Female choice selects for extreme tail length in a widowbird" published in *Nature* (1982).[146] The study showed that males with artificially elongated tails got more mates and offspring than control males, thus providing early evidence that female choice could work in nature. Likewise, herpetologist Mike Ryan (1980; see Chap. 5) had done the same in his frog experiments. These studies, the theoretical models together with the empirical evidence, contributed to putting an end to the controversy over female choice as a mechanism of sexual selection over a century after it began. Through the acceptance of female choice, females became acknowledged as an active force selecting for male traits, just as Charles Darwin had believed. Yet, just as in Darwin's writings, stereotypic assumptions about females as coy, passive and monogamous still prevailed.[147]

---

[144] Lande, 1981.
[145] Kirkpatrick, 1982.
[146] Andersson, 1982.
[147] Hrdy, 1986.

After the female choice controversy was resolved[148] both theoretically and empirically, some controversy around female choice remained, namely *why* females were choosy—whether it was for genetic or direct benefits—and this question is still to a certain extent controversial, notes Andersson.[149]

## State-of-the-Art in Sexual Selection in 1994

By 1994, sexual selection had grown into a productive and expansive research field and Malte Andersson synthesized its state-of-the-art in his classic book *Sexual selection*.[150] By then, he tells me in the interview, he had rewritten it twice, updating chapter by chapter, because the field was expanding so quickly that once the manuscript was ready, it had already become outdated. Since then, he has repeatedly declined the offer to write a follow-up, since he considers it to be impossible to keep up with the quick expansion of research in all the sub-fields.

*Sexual Selection* gives an overview of empirical studies conducted until 1990,[151] summarized in a table of studies of sexual selection mechanisms and affected characteristics. In these studies, the study species that dominate are insects, fishes, anurans and birds. Andersson points out that some traits can easily be measured, such as body size, and that calls are easier to "quantify, manipulate and synthesize than are visual or chemical signals", which may explain why these dominated sexual selection studies.[152] Among those studies, female choice was the most common mechanism (167 studies), and Andersson argues that it has been the most studied aspect because it was the most controversial one, and thereafter follows male competition (58) and male choice (30). Among the female

[148] Interestingly, population geneticists had shown experimentally that females did distinguish between different males (could identify males of their own species and preferred some males of their own species to others) by the end of the 1960s, and had ended their controversy over whether females could choose (Milam, 2010: 132). Thus, there is a parallel history of an ending of the female choice controversy in population genetics.

[149] Malte Andersson interview 2017.

[150] Andersson, 1994.

[151] Ibid: Chap. 6.

[152] Ibid: 129.

choice studies, the selected trait was most often male song or other display (72), followed by male body size (29), male visual ornaments (30), male territory (23) and other male material resources (17). Although female choice in a sense means that females have agency in the selection process, these studies were mainly focused on how males and their traits evolve. In contrast, studies on how female traits evolve included thirty studies of male choice and two studies of female-female competition, all mainly selecting for large female size. When I ask him, Andersson agrees that the field was male-biased, but he doesn't think it has much to do with gender bias in our thinking, rather that male ornaments are "awfully much more eye-catching", and that they "scream for an answer".[153] Furthermore, extreme traits are easier to work with and they also demand an explanation more urgently than subtle traits often found in females. "That's why they were studied first, not because it was specifically *males*, if the females had had those",[154] suggests Andersson, studies would have focused on females instead. Also, recounts Andersson, the opposite pattern of female competition had been observed early on, in for example jacanas (such as the wattled jacana, *Jacana jacana*), birds in which females are more ornamented and larger than males. Nevertheless, the prevalent male precedence—research starts with male-centered investigations or explanations and later includes the female-centered equivalents—in sexual selection, primarily studying evolution of male traits and their effect on reproductive success is a kind of androcentrism.[155] Although biological patterns are an important factor, the continued androcentrism in research has led to female traits becoming overlooked.[156]

Andersson's *Sexual Selection* shows the state-of-the-art of the field at that time. For example, sperm competition is dealt with in great detail, then in a passage, Andersson cites Thornhill's paper on cryptic female choice (females influencing which sperm fertilizes their eggs during or after copulation; see Chap. 6). However, he didn't mention that specific

---

[153] Malte Andersson interview 2017.

[154] Malte Andersson interview 2017.

[155] Ah-King 2022.

[156] For example female nuptial coloration in fishes (McLennan, 1994) and female birdsong (Odom et al., 2014).

term in the book, and in my interview, he comments that by then he didn't know that it would become so important.

Interestingly from an epistemology of ignorance perspective, in the section about different critiques of sexual selection,[157] feminist critique is not mentioned at all. Although in other parts of the book, empirical studies by feminist biologists (e.g., Sarah Blaffer Hrdy and Patricia Gowaty) are cited. Thus, perhaps empirical studies have been an efficient route for feminist biologists' work to get recognition in the field, more so than theoretical contributions pointing out androcentrism. However, as described in Chap. 2, feminist primatologist Jeanne Altmann's successfully convinced researchers in animal behavior to reform their observation methods.

Furthermore, *Sexual selection* is also a time document of the role of women in science. In the acknowledgments to *Sexual selection*, Andersson thanks his international and national all-male colleagues for comments on earlier drafts and a woman for preparing the reference list of the book as well as three female relatives for helping with the index. This male bias gives an illustration of the role of women and men in sexual selection or academia in Sweden and internationally at the time, from the perspective of Andersson's network.

Since *Sexual Selection,* the field has expanded with work on for example parental roles, says Andersson. He thinks people have been paying more attention to that because much more women have entered science, in ecology as well as in evolutionary biology. Here he connects the entrance of women with a shift in research questions towards including parental care, similar to the shifts in primatology discussed in Chap. 2.

## Subsequent Research on Female Ornaments

There was a lot of research on male ornaments and then, Malte Andersson asserts, "when the dirty work was done", researchers started to ask whether females had ornaments too. That was an overlooked issue, recalls Andersson and recounts the subsequent studies of female ornaments in

[157] Andersson, 1994: 17ff.

crested auklets, *Aethia cristatella*, butterflies and pipefish. Female ornaments turned out to be important too, although they are more subtle and not as conspicuous as male ornaments. Focusing on male ornaments first, and then studying female ornaments is another example of male precedence in sexual selection. Although, as Andersson points out, because male ornaments are generally more developed and glaring compared to the more subtle female ornaments, this biological phenomenon rather than cultural gender bias is the likely cause in this particular case.

However, investigations of female traits have repeatedly questioned classical sexual selection assumptions. For example, in a recent paper (2020), American ornithologist Casey Haines and collaborators call attention to that female ornamentation and female birdsong have been generally ignored until lately.[158] They emphasize the disproportionate role that women (compared to men) have played in studying female birdsong, which have led to a paradigm shift in perceptions of bird song—from traditionally being considered a sexually selected male trait, to conceiving it as ancestral in both females and males. Thus, similar to the pattern of women's role in the development of gender-reflective approaches in primatology, women ornithologists revised perceptions about female birds. This suggests that androcentrism has contributed to the maintenance of ignorance about female ornaments (Fig. 3.2).

## Lack of Resistance to Female Choice

Despite the long history of resistance to accepting female choice as a mechanism of sexual selection, Malte Andersson recounts that he did not encounter any resistance to his female choice study. Once the theoretical and experimental evidence for female choice were established, he reasons, nobody objected. He met much more resistance to his later research on female common eiders (*Somateria mollissima*) laying eggs in each other's nests, bringing clutches together and collaborating in caring for them, and the suggestion that kinship selection could be involved. There has been a lot of skepticism and challenges, and they still exist. He

[158] Haines et al., 2020.

**Fig. 3.2** Illustrating study species in Chap. 3. A) Malte Andersson's study species long-tailed widowbirds, *Euplectes progne*, (Mohanr53 at English Wikipedia, public domain). B) A pair of crested auklets, *Aethia cristatella*, in which female as well as male ornaments were studied, photo by Alaska Maritime National Wildlife Refuge (public domain)

exemplifies with the common statement that "isn't it obvious that they should avoid parasitizing close relatives?" Andersson argues that in the case of female choice studies, they were in concert with the social development in society, the women's movement was active in the 1970s, and therefore it wasn't resisted. But when Edward Wilson suggested that much of our social behavior has a genetic component and is dependent on kinship, he met with tremendous resistance, resulting in the earlier-mentioned sociobiology controversy.

Andersson's motivation for doing research was "a wonder of oddities and spectacular things" he had observed before. At the time, there were a lot of behavioral bird field studies going on, particularly in the Anglosaxian world. Birds were close at hand, because just like us, they use sight and hearing as their most important senses, so we can understand them, says Andersson. In contrast, mammals are much harder for humans to understand because they rely so much on their olfactory sense. Fishes also use visual signals and are easy to use in experiments, therefore they've also been common model species in sexual selection. Hence, Andersson describes how our human physiological senses influence which animals we focus our research on, and how our senses are forming our situated knowledges.

Andersson's situated knowledges about sexual selection were formed by both his work with specific birds, his mathematical modeling and current theoretical and empirical influences. Thus, his work, by establishing female choice as a mechanism of sexual selection, contributed to the ending of the female choice controversy.

## The Modern Development of Sexual Selection

Sexual selection has changed since Charles Darwin's time. As Evelleen Richards points out, Darwin's understanding of sexual selection differs from that of contemporary evolutionary biologists.[159] Darwin based his specific understanding on embryogenesis (that the male was the more developed sex), of animal females having an aesthetic taste for the beautiful and his particular ideas of inheritance (that traits that appear in either sex late in life develop in the same sex alone, while traits that appear early in life usually develop in both sexes[160]). In current sexual selection, female choice is usually seen as optimizing fitness (that behavior is selected which leads to enhancement of reproductive success) or an effect of sensory exploitation (that the brain through the way it works is biased toward preferring certain signals, see chap. 5), furthermore genetics have revolutionized our understanding of inheritance. Since Darwin, consensus of sexual selection has developed and incorporated Angus Bateman's experiment and Robert Trivers' Parental Investment theory into what psychologist and historian Don Dewsbury has denoted the Darwin-Bateman-paradigm,[161] which is now the prevailing theoretical framework. Dewsbury identifies the paradigm's basic assumptions as 1) males' reproductive success is more variable than females', 2) males gain more in reproductive success by mating with multiple partners than females

[159] Richards, 2017: 535.

[160] Ibid: 284.

[161] Dewsbury, 2005.

and 3) males are generally more eager to mate and indiscriminate in mating while females are discriminating and less eager.[162] Dewsbury's analysis demonstrates that gender stereotypes influenced these assumptions, and that the development of the field shows that although those assumptions probably hold in a majority of cases—in light of the amounting exceptions it is important to reevaluate them.[163]

## An Expanding Field of Study

Since the revival of sexual selection, the field has expanded from on average 29 papers per year in the 1980s, to 960 per year in the 2010s. Controversies and negotiations about sexual selection have continued; importantly there is not one agreed-upon model of sexual selection—there are many whereof some have conflicting assumptions[164] (see also the discussion on definitions of sexual selection in Chap. 7). Both the history of the eclipse and the many models of sexual selection illustrate that, in science historian Evelyn Fox Keller's words, science is a polyphonic chorus, "Scientists make up many communities, and these communities vary by subject, by methodology, by place, and by degree of influence".[165] Thus, controversies and disagreements among the multitude of approaches within the sexual selection field continue.

# Zuleyma Tang-Martínez—Criticizing the Darwin-Bateman Paradigm

Zuleyma Tang-Martínez is a rodent researcher and a prominent critic of the Darwin-Bateman paradigm, whom I met in St. Louis for an interview. She is a Venezuelan American Professor Emerita in Biology at the University of Missouri, St. Louis and the Animal Behavior Society historian. She gained her PhD in Zoology from Berkeley in 1974. She says she

[162] Ibid: 831.
[163] Ibid: 836.
[164] Richards, 2017: 533.
[165] Keller, 1983: 174.

has always had a strong interest in philosophy and history of ideas in biology.[166] During Tang-Martínez' graduate work at Berkeley, feminist primatologist Thelma Rowell—who due to antinepotism rules (whereby universities refused to hire more than one person from the same family) was given a teaching role—had a huge impact on Tang-Martínez in making her aware of the unequal treatment of women, the focus on males in animal behavior studies and that females were overlooked. Rowell used to say "don't accept something just because you read it in a textbook", she constantly encouraged questioning and demands for evidence.

A turning point for Tang-Martínez was the discovery of widespread extra-pair paternity in the early 1980s—the immediate male-focused "rape" explanations and the subsequent revelation that females could seek them out (see chap. 4)—"because that was so totally contrary to everything that I had been taught and everything that I had believed to that point". It raised questions about what else researchers had gotten wrong about this paradigm.

Another important influence on Tang-Martínez' thinking about science and paradigm shifts was science philosopher Thomas Kuhn's *The Structure of Scientific Revolutions*.[167] She had been thinking about these things for a long time but didn't start writing about them until later in her career, when she had a reputation in science and felt safer to challenge these prevailing ideas. She wanted people to know that she is "a bonafide scientist, biologist, animal behaviorist", because it makes her criticism "more palatable". She says she's very careful in her writing not to be too polemical. She has met a lot of resistance and still thinks that there remains an incredible resistance in some quarters, because there is a whole body of literature that has depended on Bateman's framework for such a long time.

In addition, Tang-Martínez recounts that her involvement with the strong women's studies department in St. Louis made her identify as a feminist. She got engaged in a reading group there and started teaching courses on women, gender and science in which she included feminist critiques and criticism of biological determinism. Engaging in these

---

[166] Zuleyma Tang-Martínez interview 2017.

[167] Kuhn, 1962.

issues had a big impact on her thinking and she talks about her knowledge as being situated by her being a woman. "I want to be thought of as a scientist, but I'm also a woman. I can't separate those two. And I very much believe that the perspectives—which is just something else in terms of how you get to where you are—I mean for me the fact that I'm a woman, very much impacts the sorts of questions that I ask, the things I think are interesting." But she finds it common among women scientists not to make that connection, and that "there's a fear that if you identify with the feminists, as a feminist, that that's going to decrease your credibility as a scientist."

In 2000, she published her first criticism of the Darwin-Bateman paradigm,[168] and in 2004 she arranged the symposium *Bateman's Principle: Is It Time for a Re-evaluation?* at the Annual Meeting of the Society for Integrative and Comparative Biology. In the resulting special journal issue published in 2005, she and ornithologist Brant Ryder analyzed the history of the Darwin-Bateman paradigm, and concluded that stereotypic perceptions about females and males have hindered and delayed the exploration of a number of research issues. The assumption of coy and monogamously mating females delayed the investigation of female multiple mating; the assumption of indiscriminately mating males with unlimited sperm supplies delayed the study of male mate choice and sperm limitation; and the assumption that females in lekking systems (in which males gather to compete and display for females) mate with the "best" male hindered the discovery of female multiple mating in these systems as well.[169] Thereby they show how this theoretical framework has produced ignorance about females.

In 2016, Tang-Martínez reviewed the history of Bateman's ideas as well as the empirical evidence for and against Bateman's assumptions in "Rethinking Bateman's Principles: Challenging persistent myths of sexually reluctant females and promiscuous males"[170] and in 2020, she edited a special issue in *Animal Behaviour* on historical approaches on animal behavior research—and co-wrote a paper about the history of the overlooked contributions of animal behavior researchers from Latin

[168] Tang-Martínez, 2000.

[169] Tang-Martínez & Ryder, 2005: 821.

[170] Tang-Martínez, 2016.

America.[171] Furthermore, as the first Latinx president of the Animal Behavior Society, she's been championing diversity in the field.[172] These criticisms and interventions illustrate Tang-Martínez' engagement with how history, philosophy and sociology of science intersect with the study of animal behavior—and who is doing the science forms the knowledge production.

Tang-Martínez situated knowledges about sexual selection were formed by her feminist insights from being a minority woman in science as well as engaged in the gender and science literature, her historical approach and broad scientific reviews of current empirical studies. Thereby she has contributed to illuminating the production of ignorance about females resulting from the theoretical framework of the Darwin-Bateman paradigm.

## Chapter Conclusions

This chapter concerns the turning appreciation of female choice in the controversy over female choice as a mechanism of sexual selection, which proceeded from Charles Darwin's time until around 1980. During this preamble to the female turn, perceptions about females as coy, passive and monogamous predominated. The controversy started when Darwin's contemporary scientists did not accept female choice among animals, leading to the disregard and slumbering existence of sexual selection (apart from its application to humans). Yet, feminists at the time both criticized the male bias in sexual selection, and used sexual selection and female choice as arguments for women's social and reproductive autonomy. As they were simultaneously excluded by the emerging scientific professional establishment, their calls for studying female special characteristics and taking women's experiences into account were ignored in science.

As scientific interest in sexual selection surged in the 1970s, whether female choice could function as a mechanism of sexual selection became an urgent question to solve. After theoretical biologists had cast doubt

[171] Jaffe et al., 2020.

[172] Lee, 2020.

upon female choice, other theoretical and empirical biologists became even more interested in proving them wrong. Around 1980, mathematical models had shown that female choice could lead to evolution of male characteristics that were detrimental to survival, and empirical evidence of choosing females accumulated, which lead to the end of the female choice controversy in sexual selection. Yet, stereotypical notions of females as passive and coy lingered, sustaining ignorance about active and multiply mating females.

Charles Darwin's situated knowledges about sexual selection were influenced by his understanding of bird breeding, his experiences as a gentleman naturalist on a British around-the-world tour, a multitude of information gained by his many correspondents around the world, his theories of embryogenesis and inheritance, as well as Victorian societal ideals. Furthermore, his own experiences corroborated his views of women and female animals as coy and nurturing. Despite this background, he cast female animals (but not women) in the active role of choosing mates.

Alfred Russel Wallace, with his long-term experience of tropical expeditions collecting insects and butterflies, focused on the evolution of camouflage through natural selection. To him, females were the ones who had evolved their dull plumages and there was no need to explain male extravagant traits through female choice. Therefore, he largely dismissed sexual selection except for the potential of female choice in humans in his socialist egalitarian vision for societal change.

The American feminists that were engaged in sexual selection in the late nineteenth century formed their situated knowledges by their feminist insights, their own experiences and their knowledge about variability in animal sexual and domestic relationships. They claimed that male choice in humans was unnatural, since female choice of mates was the norm among animals.

Ornithologist Malte Andersson's particular view on sexual selection was formed by his studies of particular birds, his mathematical modeling and the current theoretical and empirical influences as well as his broad overview of the field resulting from his 1994 state-of-the-art book. His experimental study of manipulated male ornaments in widow birds contributed to ending the controversy over female choice as a mechanism of sexual selection.

Rodent researcher Zuleyma Tang-Martínez' situated knowledges of sexual selection resulted from feminist insights from being a minority woman in science as well as her engagement with the gender and science literature, her historical approach and broad scientific reviews of current empirical studies. Thereby she has shown how the Darwin-Bateman paradigm has produced ignorance about females (and males) through its stereotypical assumptions about females as coy, passive and monogamous.

This chapter shows that these researchers' and amateurs' situated knowledges were formed by their cultural context, particular study animals, their experiences, their specific methodology and, for some, by their feminist insights.

This history includes several examples of production of ignorance. First, that female choice as a mechanism of sexual selection was not accepted by Darwin's contemporary scientists and led to the eclipse of sexual selection. Second, that feminists who engaged in sexual selection and their gender-reflective approaches were excluded from science. Thus, their exclusion contributed to the production of ignorance about females in sexual selection. Furthermore, the border between professionals and amateurs again sustained ignorance about female animals' agency, when amateur research on sexual selection was overlooked in the beginning of the twentieth century.

Male precedence is prevalent in sexual selection research, as shown by the prevailing focus on male traits and their effects on male reproductive success in Malte Andersson's review of the field in *Sexual Selection*. Although male precedence is partly due to biological patterns of common male extraordinary ornaments, continued androcentrism has sustained ignorance about the evolution of female traits.[173]

Thus, throughout this history, ignorance about females has been produced—through the ignorance of female choice as sexual selection, male precedence and continued androcentrism of sexual selection theory, and the exclusion of women, overlooking amateurs' findings and disregard of gender-reflective approaches in science.

---

[173] Ah-King 2022.

## References

Ah-King, M. (2022). The history of sexual selection research provides insights as to why females are still understudied. Nature Communications 13(1) 6976 10.1038/s41467-022-34770-z

Andersson, Malte. (1975). *Some aspects of behaviour and ecology in skuas (Stercorariinae).* Dissertation. Gothenburg: Gothenburg University.

Andersson, M. (1982). Female choice selects for extreme tail length in a widowbird. *Nature, 299*, 818–820.

Andersson, M. (1994). *Sexual selection.* Princeton University Press.

Bateman, A. J. (1948). Intra-sexual selection in *Drosophila. Heredity, 2*, 349–368.

Birkhead, T., Wimpenny, J., & Montgomerie, R. (2014). *Ten thousand birds: Ornithology since Darwin.* Princeton University Press.

Blackwell, Antoinette Brown. (1875). *The Sexes throughout Nature.* New York: G. P. Putnam's Sons. Reprinted through the Pioneers of the Woman's Movement series, Westport: Hyperion Press, 1976. (Indicated page numbers refer to the reprinted edition.)

Blackwell, Antoinette Brown. (1876). The Savants of the Woman Question. *Woman's Journal,* July 22:1. (Cited in Hamlin 2014: 62.)

Campbell, B. C. (Ed.). (1972). *Sexual selection and the descent of man: 1871–1971.* Aldine Publishing Co.

Clarke, E. (1873). *Sex and education; or, a fair chance for the girls.* James R. Osgood.

Cronin, H. (1991). *The ant and the peacock: Altruism and sexual selection from Darwin to today.* Cambridge University Press.

Darwin, Charles. (1871). Republished in 1981. *The Descent of Man and Selection in Relation to Sex.* Reprint of the 1871 edition published by J. Murray, London. Princeton University Press.

Dawkins, Richard. (1976). Republished in 2016. *The selfish gene.* 40th anniversary edition. Oxford, United Kingdom; Oxford University Press.

Desmond, A., & Moore, J. (2009). *Darwin's sacred cause: Race, slavery, and the quest for human origins.* University of Chicago Press.

Dewsbury, D. (2005). The Darwin-Bateman paradigm in historical context. *Integrative and Comparative Biology, 45*, 831–837.

Gieryn, T. F. (1999). *Cultural boundaries of science: Credibility on the line.* University of Chicago Press.

Gowaty, P. A. (2003). Sexual natures: How feminism changed evolutionary biology. *Signs, 28*(3), 901–921.

Haines, C. D., Rose, E. M., Odom, K. J., & Omland, K. E. (2020). The role of diversity in science: A case study of women advancing female birdsong research. *Animal Behaviour, 168*, 19–24. https://doi.org/10.1016/j.anbehav.2020.07.021

Hamlin, K. A. (2014). *From eve to evolution: Darwin, science, and women's rights in gilded age America*. The University of Chicago Press.

Hamlin, K. A. (2021). Darwin's bawdy: The popular, gendered and radical reception of the descent of man in the US, 1871–1910. *British Journal of the History of Science Themes, 6*, 115–131. https://doi.org/10.1017/bjt.2021.6

Haraway, D. J. (1988). Situated knowledges: The science question in feminism and the privilege of partial perspectives. *Feminist Studies, 14*(3), 575–599.

Haraway, D. J. (1989). *Primate visions: Gender, race and nature in the world of modern science*. Routledge.

Hrdy, S. B. (1986). Empathy, polyandry, and the myth of the coy female. In R. Bleier (Ed.), *Feminist approaches to science*. Pergamon Press.

Hrdy, S. B. (2000). *Mother nature: A history of mothers, infants, and natural selection*. Pantheon, cop.

Jaffe, K., Correa, J. C., & Tang-Martínez, Z. (2020). Ethology and animal behaviour in Latin America. *Animal Behaviour, 164*, 281e291.

Keller, E. F. (1983). *A feeling for the organism: The life and work of Barbara McClintock*. W. H. Freeman.

Kirkpatrick, M. (1982). Sexual selection and the evolution of female choice. *Evolution, 36*(1), 1–12. https://doi.org/10.1111/j.1558-5646.1982.tb05003.x

Kuhn, T. S. (1962). *The structure of scientific revolutions*. University of Chicago Press.

Lack, D. (1968). *Ecological adaptations for breeding in birds*. Methuen Ltd.

Lande, R. (1981). Models of speciation by sexual selection on polygenic traits. *Proceedings of the National Academy of Science USA, 78*(6), 3721–3725.

Lee, D. (2020). Diversity and inclusion activisms in animal behaviour and the ABS: A historical view from the U.S.a. *Animal Behaviour, 164*, 273e280.

McLennan, D. A. (1994). Changes in female colour across the ovulatory cycle in the brook stickleback, Culaea inconstans (Kirtland). *Canadian Journal of Zoology, 72*(1), 144–153. https://doi.org/10.1139/z94-018

Milam, E. L. (2010). *Looking for a few good males: Female choice in evolutionary biology*. John Hopkins University Press.

Mill, J. S. (1859). *On liberty*. John W. Parker & Son.

Mill, J. S. (1869). *The subjugation of women*. Longmans, Green, Reader, and Dyer.

Morgan, T. H. (1903). *Evolution and adaptation*. Macmillan.

O'Donald, P. (1967). A general model sexual and natural selection. *Heredity, 22*, 499.
Odom, K. J., Hall, M. L., Riebel, K., Omland, K. E., & Langmore, N. E. (2014). Female song is widespread and ancestral in song birds. *Nature Communications, 5*, 3379. https://doi.org/10.1038/ncomms4379
Parker, G. A. (1970). Sperm competition and its evolutionary consequences in the insects. *Biological Reviews, 45*, 525–567.
Richards, E. (2017). *Darwin and the making of sexual selection*. The University of Chicago Press.
Richardson, S. S. (2013). *Sex itself: The search for male and female in the human genome*. University of Chicago press.
Russett, C. E. (1989). *Sexual science: The Victorian construction of womanhood*. Harvard University Press.
Segerstråle, U. C. O. (2000). *Defenders of the truth: The battle for science in the sociobiology debate and beyond*. Oxford University Press.
Simmons, Leigh W. (2001). *Sperm competition and its evolutionary consequences in the insects*. Monographs in behavior and ecology. Princeton N.J; Princeton University Press.
Smith, J. M. (1991). Theories of sexual selection. *Trends in Ecology and Evolution, 6*(5), 146.
Tang-Martínez, Z. (2000). Paradigms and primates: Bateman's principle, passive females, and perspectives from other taxa. In S. C. Strum & L. M. Fedigan (Eds.), *Primate encounters: Models of science, gender and society* (pp. 261–274). University of Chicago Press.
Tang-Martínez, Z. (2016). Rethinking Bateman's principles: Challenging persistent myths of sexually reluctant females and promiscuous males. *Journal of Sex Research, 53*(4–5), 532–559. https://doi.org/10.1080/00224499.2016.1150938
Tang-Martínez, Z., & Ryder, B. T. (2005). The problem with paradigms: Bateman's worldview as a case study. *Integrative and Comparative Biology, 45*, 821–830.
Trivers, R. L. (1972). Parental investment and sexual selection. In B. Campbell (Ed.), *Sexual selection and the descent of man* (pp. 136–179). Aldine.
Williams, G. C. (1966). *Adaptation and natural selection: A critique of some current evolutionary thought*. Princeton University Press.
Wilson, E. O. (1975). *Sociobiology: The new synthesis*. Belknap Press of Harvard University Press.
Zahavi, A. (1975). Mate selection—A selection for a handicap. *Journal of Theoretical Biology, 53*(1), 205–214. https://doi.org/10.1016/0022-5193(75)90111-3

# 4

# A Female Turn in Bird Research

The arrival of red-winged blackbirds, *Agelaius phoeniceus,* with their glossy black plumages, shiny red shoulder patches and their classic "conk-la-ree" song is a sign of spring for many North Americans.[1] But red-wings can also be a pest as they occur in large numbers and feed on farmers' crops. In 1971, biologist Olin Bray and his colleagues[2] in Colorado conducted an experiment on wild red-wings with the intention to find a way to reduce the flocks—without upsetting the public by shooting large numbers of the popular birds. They knew from earlier studies that males held territories with a "harem"[3] of one to four females ("harem" was commonly used anthropomorphizing terminology at the time to describe a mating system in which a male mates with several females), and therefore they suggested that sterilizing the territorial males should do the trick of preventing reproduction. Thus, the team surgically sterilized territorial males, collected the eggs from their territories and hatched them in an artificial incubator. In contrast to the investigators' predictions, 69 percent of the eggs in sterilized males' nests were fertile. They suggested

[1] Zuk, 2002:64.

[2] Bray et al., 1975.

[3] Ibid: 187.



M. Ah-King, *The Female Turn*, https://doi.org/10.1007/978-981-19-7161-7_4

several possible explanations for this unexpected finding—females keeping sperm alive since before the sterilization of their male partners, new females with live sperm moving into the territories (they did not mark the females, so there was no way of knowing whether the egg-layers were the original females or new ones), or "promiscuity by females"[4] (promiscuity is another anthropomorphic term). The authors propose that the most likely explanation is "female promiscuity", because of the pattern of infertile eggs—the sterilized males that had territories furthest away from fertile males had the lowest percentage of fertile clutches. This "female promiscuity" went against prevailing assumptions about females and pioneered the unmaking of ignorance about female multiple mating in birds. Bray's study was not motivated by feminist incentives to investigate female birds' lives as some primate studies discussed in Chap. 2; instead, their discovery was an accidental by-product of their attempt to cull red-wings. Although pioneering, Bray's study was only occasionally cited during the following years, and the topic of female multiple mating in birds came into the limelight of sexual selection research many years later.

This chapter concerns shifting perceptions about females in bird research—a female turn in the subfield of sexual selection concerning birds. I start out by discussing why birds have been of special interest in sexual selection research, then describe and explain the prevailing assumption of female birds as monogamous, the technological innovations which enabled a shift in knowledge, the continued assumptions of passive females and how these have, at least partly, shifted toward acknowledging active females. Thus, I first present a general analysis of shifting perceptions in bird research, and thereafter recount the perspectives and experiences of two ornithologists that I interviewed, Patricia Gowaty and Simon Griffith. Furthermore, I discuss the influence of the researchers' study species on their situated knowledges of females and sexual selection.

Throughout the chapter, I investigate how the researchers' situated knowledges[5] and ignorance were formed to acknowledge active and multiple mating females, or not. By exploring who knew about active and multiple mating females, and who did not, as well as why, the analysis

---

[4] Ibid: 191.

[5] Haraway, 1988.

shows how ignorance about females was produced. Furthermore, this history also shows how ignorance has been unmade through certain researchers' strategies to incorporate feminist critique and gender-reflective approaches[6] into this field, but also that production of ignorance continues.

## "The Polyandry Revolution"

One account of the shift in perceptions about females is provided by renowned evolutionary theoretician and entomologist Geoff Parker and noted bird researcher Tim Birkhead (2013) in their historical review of what they call the "Polyandry Revolution" (polyandry refers to a female mating with several males):

> …there has undoubtedly been a major and important shift in how we view female sexual behaviour and polyandry [in animals]: while most work prior to around 1980 tended to ignore female interests, there is now a much greater focus on the female viewpoint in both pre- and post-copulatory aspects of mate choice, and the adaptive function of female behaviour and morphology.[7]

Parker and Birkhead describe a general shift away from observational and explanatory male bias toward including female perspectives[8] in sexual selection research and exemplify this shift with a historical account of the development in bird research. However, in my interviews, I found that several investigators did not agree with the description of the "polyandry revolution" as a general phenomenon in the wider community of evolutionary biologists. As I will get back to in Chap. 5, especially the entomologists did not agree to a "polyandry revolution" in insect research, since they knew about widespread polyandry among insects already in the 1950s. My interpretation is that ornithology (bird research) was the

[6] Richardson, 2013: 17.

[7] Parker & Birkhead, 2013: 2.

[8] "Female perspectives" is an often-used term in evolutionary biology meaning to consider the world from a female point of view, as opposed to an androcentric perspective.

primary scene for the "polyandry revolution" of shifting perceptions about female multiple mating. However, it is questionable whether it should be called a revolution, as there is no consensus among historians and philosophers about how scientific revolutions should be defined.[9] Still, biologists generally acknowledge that a shift in perceptions has occurred from assumptions of females as generally monogamous to the recognition of widespread polyandry.[10] Thus, perceptions about females in bird research have shifted from principally monogamous to almost ubiquitously mating with multiple males, and also from passive to including opportunities for active sexual strategies and aggression—as well as from expecting females to be uniform (in their preferences of males and in their reproductive success) to acknowledging variability among females.[11] These shifts represent different aspects of the still ongoing process of the female turn in bird research.

## Classical Study Species

In sexual selection, birds have been the classical group to study and are therefore often described as the paradigmatic animals in this field.[12] Bird examples were prominent in Charles Darwin's book on sexual selection (four chapters and over 200 pages concerned birds), and male ornaments and birdsong were central to his ideas.[13] Birds are also readily available to study, they can be found everywhere and they provide ample possibilities to make experiments. In contrast to western primatology, in which field studies are expensive and dependent on institutional support,[14] studies of birds can be performed in any country and behavioral studies are inexpensive. When I interviewed ornithologist Malte Andersson,[15] whose work on

[9] Nickles, 2017.
[10] Birkhead, 2000; Taylor et al., 2014.
[11] Birkhead, 2000; Gowaty, 1997a, Griffith et al., 2002.
[12] For example Ryan, 2010: 474.
[13] Richards, 2017:433ff.
[14] Haraway, 1986:79.
[15] Malte Andersson interview 2017.

female choice for elaborate male ornaments is described in Chap. 3, he noted that there was a lot of field research on birds from the UK and the US, and mentioned an inspiring bird behavior book from the UK published in 1935. He thinks the abundance of research on birds has to do with "our senses being similar". For birds, just as for humans, he says, "vision and sound are important senses, so we can understand them, in contrast to our understanding of mammals, that are mainly oriented by smell." Andersson describes how our human physiological senses enable, steer and limit our knowledge making, in other words, he illustrates different aspects of how biologists' knowledges are situated. As we shall see later in this chapter, there are many other aspects of how researchers' knowledges are situated, through their specific study species, through their theoretical framework, their feminist insights and gender-reflective approaches.

Historically, ornithology has a long tradition of consisting almost entirely of men. In the nineteenth century, the way bird studies were conducted—by walking around in the countryside and shooting birds, climbing trees and cliffs to collect eggs—were not considered appropriate for a woman, neither were women allowed in the collecting networks in the US.[16] In the beginning of the twentieth century, some formal barriers were removed, but discrimination was abundant. Very few women pursued a career in ornithology before the 1970s.[17] Robert Montgomerie, the chair of the history committee at the American Ornithological Society, reports that these women's contributions have been relatively invisible.[18] This male dominance among researchers in ornithology is historically correlated with gender bias in the content of science, just as in western primatology (see Chap. 2) and sexual selection in general (Chaps. 3 and 7). As we shall see, the female turn is an effect of negotiations of knowledge—that researchers challenged the prevailing perceptions about females as coy, passive and mating with only one male.

---

[16] Barrow, 1998:43.

[17] Montgomerie, 2019.

[18] Montgomerie has during 2018–2019 published a number of blog posts for the American Ornithological Society about overlooked women ornithologists (Montgomerie 2018, 2019).

## Assumptions of Monogamous Females

The assumption of monogamous females was based on natural history observations of birds as well as sexual selection theory's assumption that females would not benefit from mating with more than one male. The renowned British ornithologist David Lack reviewed breeding systems in birds and concluded that over 90 percent of the over 9000 bird species are monogamous, maintain strong pair-bonds and often show biparental (both male and female) care.[19] According to Lack, birds are monogamous because the female is unable to bring up the offspring by herself. He therefore assumed females to mate only with their social mate, the partner with whom they pair-bond and care for offspring. These assumptions in combination with Charles Darwin's assumptions of "coy" and "discriminatingly passive" females and Bateman's proposition that females do not benefit from mating with multiple males (see Chap. 3), meant that the assumption of monogamous females prevailed.

Likewise, the presumption of females being twosome and faithful was closely linked to the assumption of females as monogamous, especially among birds—they were, in herpetologist Mike Ryan's words, "the icons of morality for a long time of not fooling around outside of the pairbond".[20]

Malte Andersson recalls that except for a few socially polyandrous species like *Jacanas*, females mating with several males was not generally acknowledged before it became possible to make molecular analyses of parentage.[21] He says that prior to that technological innovation, people did speculate about the possibility of female multiple mating and tried to make observations. Nevertheless, a literature review of sperm competition (competition among different males' sperm for fertilization of eggs) in 1984 reviewed behavioral evidence of extra-pair copulations (mating with individuals outside of their social pair) in 104 species of birds.[22] The review authors note that at that time, there were few studies that provided

[19] Lack, 1968.

[20] Mike Ryan interview 2017.

[21] Malte Andersson interview 2017.

[22] McKinney et al., 1984.

direct evidence of multiple paternity and sperm competition in birds. However, that was about to change.

## Technical Innovations

New technologies have been important for changing perceptions about females, because they enabled testing of hypotheses that were difficult to explore before. They revealed that female multiple mating was much more widespread than predicted, especially in supposedly monogamous birds. In the late 1970s–1980s, it was cumbersome to find evidence of parentage. Early methods included analyzing protein variation in allozymes (different variants of the same protein) by electrophoresis (a laboratory technique to separate, for example, proteins by their size and electrical charge). It was a strenuous method. It did not work for all species and also did not serve to identify a particular individual as parent, but could only be used for parentage exclusion.[23] The other methods used at this time were tracing inheritance of plumage color—as it is possible to trace inheritance of eye color—and estimate heritability of certain morphological traits, such as by comparing the tarsus[24] bone length with prospective mothers and fathers.[25] These methods also could not definitely prove parentage, but merely exclude certain individuals as parents.

In contrast, DNA fingerprinting enables distinguishing one individual from another (with the exception of identical twins), and therefore this technology enabled demonstrating genetic relationships and more precisely test hypotheses about sexual interactions. Early on the technique was a bit of a bottleneck for the researchers, because although the methods were available, they were not easy to use. "Some laboratories became very skillful, for example Terry Burke at the University of Sheffield, he was involved in a huge amount of studies, because he had the equipment and the skills" recalls behavioral ecologist Charlotta Kvarnemo.[26]

[23] Zuk, 2002:65–66.

[24] The tarsus is the part of the foot between the heel and the ball of the foot.

[25] Griffith et al., 2002: 2196.

[26] Charlotta Kvarnemo interview 2017.

In 1987, Terry Burke and Michael Bruford, both molecular ecologists, published the first paper using DNA fingerprinting in birds (on house sparrows, *Passer domesticus*) to test parentage, foretelling that the method would be very useful for studying sexual selection and mating behavior in wild birds and mammals.[27] Subsequently, a long series of papers demonstrated that birds often have clutches with multiple paternity. In the 2000s, the consensus was reached that extra-pair paternity is widespread among birds, even among presumedly monogamous species, although the level of extra-pair paternity varies considerably between and within species.[28] Thus, DNA fingerprinting led to the shift from perceiving bird females as principally monogamous (with the exception of the few species that were already known to be polyandrous), to acknowledging widespread female multiple mating among birds. However, as we shall see, this discovery initially did not change the prevailing perception of females as passive.

Radio-tracking was another important technology, it was used to track animals in the wild as early as the 1960s, but the transmitters were initially expensive, short-range and large, only after the development of new technologies—geolocators, GPS trackers, satellite and ground receivers—could it be used for small birds too.[29] As we shall see, ornithologist Susan Smith used radio-tracking to follow female black-capped chickadees (*Poecile atricapillus*) and could thereby show that females actively seek extra-pair copulations. Thus, technological innovations may open up possibilities to investigate questions that had not been possible to test before, yet someone also needs to ask those new questions and have the motivation to pursue them.

## Assumptions of Passive Females

Prevailing sexual selection theory assumed females to be passive in sexual interactions. As DNA analyses of parentage accumulated, the extent of female multiple mating was established as much more widespread than

[27] Burke & Bruford, 1987.

[28] Griffith et al., 2002.

[29] Birkhead & Montgomery, 2020: 2.

expected, especially among presumably monogamous birds. However, perceptions of passive females lingered, therefore initial explanations for female multiple mating tended to be that males were coercing or harassing females into mating with them.[30] Thus, early research interpreted widespread female extra-pair mating as a result of male strategies.[31] Many investigators were convinced that males alone were causing extra-pair mating, because they assumed females would not seek them, as theory predicted that females had no reason to mate with more than one male. For example, Tim Birkhead, in a 1987 review of sperm competition in birds, presents extra-pair copulations as something males do to females:[32] males are described as actively timing copulations when females are fertile, and at the same time guarding their own mates or copulating frequently with the latter in order not to become "cuckolded" themselves (the anthropomorphic term "cuckold" is a derogative term for a man whose wife is unfaithful). Hence, females are depicted as passive or reluctant recipients of extra-pair copulations. This view, that selection on males was shaping extra-pair mating strategies, prevailed until the mid-1990s.[33]

As demonstrated in Marcy Lawton et al.'s (1997) gender analysis of ornithological studies, a study on spotted sandpipers (*Actitis macularius*) provides a striking example of this androcentrism: ornithologist Lewis Oring et al. (1992) write, "Here we report that males pairing early in the season cuckold their females' later mates by means of stored sperm."[34] This study presents females as passive receptacles for the sperm of multiple males—in a polyandrous species in which females form sequential breeding alliances with up to four males (that incubate eggs and hold nesting territories) during one breeding season. The authors show that females can store sperm for a month and use it after mating sequentially with three other males, but the females are depicted as being exposed to an active male cuckolding strategy. Lawton and coworkers point out that since the

[30] Zuleyma Tang-Martínez interview 2017.

[31] Lifjeld et al., 1994:244, Burley & Parker, 1998:1, Gray, 1998:62.

[32] Birkhead, 1987:268.

[33] Burley & Parker, 1998:2.

[34] Oring et al., 1992: 631.

males are in focus, the authors never ask the interesting questions about possible mechanisms and benefits for females.[35] Thus, androcentrism produces ignorance about females, by blocking knowledge production about females, neglecting certain questions and alternative explanations, such as what could be the benefit for females of using stored sperm? Lawton and colleagues provide an alternative interpretation of the observed phenomenon: females often actively use sperm from their first chosen male, and produce many clutches of mixed paternity, and thus make very complex decisions about reproduction throughout the breeding season.[36] They suggest it is a female bet-hedging strategy in the predation-prone environment they live in,[37] an interpretation which was ignored when the investigators focused solely on male strategies.

Likewise, Lawton and colleagues analyzed Tim Birkhead's *The Magpies* (*Pica pica*) (1991) in which males are presented as the active sex in extra-pair copulations. This study too focuses on males: Birkhead describes males as taking advantage of females' sperm storage tubules and initiating extra-pair copulations. Likewise, he portrays females as passive and "belonging" to males. Furthermore, he does not discuss the possible benefits of sperm storage tubules for the females, the very sex that has developed them.[38] Yet, in Birkhead's observations, all cases of extra-pair copulations were initiated by females; still, his only discussion of benefits of extra-pair copulations is focused on males, and benefits to females were not even considered in this study. Again, the unconscious androcentric assumptions about females as belonging to their male partners, and males as the active part in mating[39] contribute to the production of ignorance about what females are doing and for what benefits. Lawton and coworkers' reinterpretation of Birkhead's findings suggests that females actively initiate extra-pair copulations with specific males that they have observed to hold a territory and invest in parental care. Hence, looking past gender bias in what they call "the mask of theory" enables new interpretations

[35] Lawton et al., 1997: 78.

[36] Ibid: 79.

[37] Ibid: 80.

[38] Ibid: 74.

[39] Ibid: 73.

that take gender perspectives into account. Lawton and colleagues reveal gender bias in several 1990s bird studies, and thus their gender analyses unmake ignorance about females and have contributed to shifting perceptions about females, from passive to having active sexual strategies.

## Casting Doubt upon Early Interpretations of Active Females

Despite that the assumptions about females as passive prevailed, there exist very early accounts of females initiating or readily accepting to mate with others than their social mate. Already in 1937, ornithologist Harry Hann reported voluntary multiple mating by female ovenbirds (*Seiurus aurocapilla*), a species of small songbird: "One female was seen copulating with a neighboring male three times, and a second neighboring male once, in the space of a few minutes. This took place in the female's home territory, and her mate was not present to object. She ate and gave the mating call between times."[40]

Hann's report is included among the 33 species with unforced extra-pair copulations cited in the aforementioned 1984 review of sperm competition in birds,[41] though the review authors make the reservation that it might be "possible that the female's willingness was apparent rather than real".[42] Despite the many accounts of sexually active females, thus casting doubt upon earlier interpretations led to ignorance about sexually active females. Such accounts as Hann's did not unsettle the prevailing view of females as passive and ignorance about active females was sustained. Hence, female extra-pair copulations continued to be interpreted as imposed by males,[43] and ignorance about female active participation continued until the late 1980s and beyond.

[40] Hann, 1937: 154.

[41] McKinney et al., 1984: 524.

[42] McKinney et al., 1984:538.

[43] For example Birkhead, 1987.

## Shifting Perceptions toward Active Females

In the 1980s, a number of challenges were presented to the prevailing view that males were causing females to mate with multiple males. Ornithologist Susan Smith revealed females' active strategies in extra-pair mating by focusing on what the females were doing. She questioned the prevalent interpretation that extra-pair copulations result from territorial males' inability to fight off other males from their territory. This interpretation was based on the implicit assumptions that females do not move beyond the boundaries of their home territory, and that females mate unselectively with any male that approaches them. Smith did not believe these assumptions and she used data from her long-term field study of radio-tracking color-marked black-capped chickadees, to test them.[44] She found that females mated selectively with extra-pair males of higher rank than their current mate, and those copulations mainly took place in the higher-ranked male's territory—suggesting that extra-pair copulations were sought out and initiated by females.[45] Thus, Smith's study rejected the implicit assumptions of passive and territory-bound females. She writes that bird copulations are very hard to observe in the field, as a copulation takes only about a minute, so it was only because she started to pay extra attention to females wandering off their territories at dawn and follow them that she was able to notice these extra-pair matings.[46] Smith's study from 1988 was a starting point for opening up for perceptions of active females in extra-pair copulation studies.[47] The shift from interpreting extra-pair matings as due to male strategies toward acknowledging female active strategies is one example of male precedence in sexual selection[48]—that research starts with male-centered investigations or explanations and later includes the female-centered equivalents, a kind of androcentrism.[49] This pattern of male precedence is repeated in many

[44] Smith, 2003: 87.

[45] Smith, 1988.

[46] Ibid: 19.

[47] Parker & Birkhead, 2013: 9.

[48] Ah-King 2022.

[49] Androcentrism is the practice of placing the male point of view in the center of one's worldview.

areas of sexual selection, in primatology (see Chap. 2), the focus on male ornaments (Chap. 3) and as we will see, in studies of genital evolution and post-copulatory sexual selection (Chap. 6).

Assumptions of males being the active sex are connected to the assumption that dominance hierarchies are most important among males. One example is the Pinyon jay (*Gymnorhinus cyanocephalus*) study from 1992, mentioned in Chap. 1, that overlooked female dominance hierarchies and dismissed female aggression as a bird version of PMS (called PBS, pre-breeding syndrome[50]). In contrast, Marcy Lawton and colleagues' gender analysis suggested that in this particular species, females are aggressive and hold breeding territories.[51] Similarly, Susan Smith demonstrated that in rufous-collared sparrows (*Zonotrichia capensis*), females are dominant to males during the breeding season, which also challenged the assumption that males are dominant to females. When she found this female breeding dominance, she reviewed earlier studies and found that it was a widespread although overlooked pattern among socially monogamous bird species.[52] In "Henpecked males—the general pattern in monogamy?" Smith (1980) points out that this female dominance was overlooked and also that ornithologist Margaret Morse Nice had challenged the assumption of male breeding dominance already in 1949. Since female dominance was overlooked in the Pinyon jay study published in 1992 as well, Smith's 1980 review was also neglected.[53] Thus, challenges to the prevailing view that males are dominant to females have been presented multiple times in bird research, but they were overlooked and seldom cited. This lack of citation is one way in which ignorance about females has been sustained.

After Smith's study on black-capped chickadees, several subsequent studies corroborated the finding that females actively seek extra-pair matings.[54] An ornithological monograph published in 1998 illustrates this

[50] Lawton et al., 1997.

[51] Lawton et al., 1997.

[52] 37 species of birds from 18 families. Smith, 1980:57.

[53] So far it has only been cited 29 times (Web of Science April 15 2020).

[54] For example, Gowaty & Bridges, 1991, Wagner, 1991, Kempenaers et al., 1992, Sheldon, 1994, Stutchbury et al. 1994, Gray, 1997, Double & Cockburn, 2000.

shift in views toward active females.[55] It is noteworthy that women researchers were often early to describe female as active: Patricia Gowaty (1985b),[56] Susan Smith (1988), Elizabeth Gray (1994),[57] Bridget Stutchbury and Diane Neudorf (1998)[58]—but also that they were not alone. A number of male researchers put forward active female birds too, for example, Eivin Røskaft (1983),[59] Richard Wagner (1991), and Bart Kempenaers[60] and colleagues (1992). However, it is interesting that among these ornithologists, I have only found some women spelling out an explicit strategy (whether they call it feminist or not) to counteract gender bias.[61] The identification of gender bias has spurred them to focus specifically on overlooked females and engage in gender-reflective approaches, as was the case with feminist primatologists in Chap. 2.

An example of an immediate shift in interpretations away from passive females occurred in ornithologist Marion Petrie's study of moorhens (*Gallinula chloropus*), in which males and females look alike, and the females are only slightly larger than the males. In my interview, Petrie recounts that she was observing moorhens and taking notes on who was fighting with whom, but it was not until she got back from the field and into the office and identified the fighting individuals with the help of her color markings that she realized that, contrary to prevailing assumptions about male competition, it was the *females* that were fighting over the males.[62] Here, just as in the reinterpretation of Pinyon jay dominance, Petrie showed that competition among females can be fierce.

Over time, the accumulation of studies demonstrating active female strategies shifted the pendulum from generally interpreting males as causing extra-pair matings toward interpreting females as generally in control

[55] Parker & Burley, 1998.

[56] Gowaty, 1985b.

[57] Gray, 1994.

[58] Stutchbury & Neudorf, 1998.

[59] Røskaft, 1983.

[60] With a mixed-sex group of coworkers Kempenaers et al., 1992.

[61] Gowaty, 1997a, Smith, 2003.

[62] Marion Petrie interview 2018. The resulting publication was entitled "Female moorhens compete for small fat males." Petrie, 1983.

of extra-pair matings.[63] For example, ornithologists Mary Fitch and Gary Shugart argued that it is unlikely that forced copulations are a successful male strategy in the majority of bird species lacking a penis[64] (only few birds have penises, for example, ducks).

This shift toward interpreting females as active was followed by a review questioning this interpretation, arguing that the widespread assumption that extra-pair paternity is a female strategy is unsupported.[65] Still, there exist rather divergent views among bird researchers. Both Patricia Gowaty and Simon Griffith, whom I interviewed, consider the perception of passive females to largely prevail in ornithology still. Yet, a number of studies do present female behavior as active and there is now a general acknowledgment that females can initiate extra-pair matings. This shift away from assumptions of universally passive females toward acknowledging opportunities for female active sexual strategies, I consider as one aspect of the female turn in ornithology.

Thus, both women and men have presented female birds as active, based on their empirical observations of specific species, but some women have used their feminist insights to pursue questions about female active strategies and thereby purposefully unmake ignorance about females in bird studies.

Similar shifts away from perceptions of females as passive have taken place in other areas of biology, for example, in fertilization biology. Anthropologist Emily Martin made a classic analysis (1991) of the scientific descriptions of sperm and eggs.[66] She found that fertilization was described through metaphors of heroic active sperm competing among each other, pursuing, penetrating and fertilizing passive eggs. Since theory predicted sperm to be active, studies were focused on measuring traits and velocity of sperm—and investigations of the role of eggs were delayed. New findings revealed that eggs play an active role in fertilization—sperm cannot penetrate eggs by the force of their speed, adhesive molecules in both the egg and the sperm bind them together, thereafter sperm release

[63] Petrie & Kempenaers, 1998, Double & Cockburn, 2000.

[64] Fitch & Shugart, 1984: 124.

[65] Westneat and Stewart, 2003: 388.

[66] Martin, 1991.

enzymes that break down the membrane. These findings have led to a shift toward acknowledging eggs as active—for example, a widely used textbook describes fertilization as the process by which egg and sperm "find each other and fuse".[67] Thus, the (incomplete) shift away from perceptions of female birds as passive has a parallel in the shift away from ideas of eggs as passive in the fertilization process.

## Patricia Gowaty—Feminist Pioneer on Polyandry, Female Aggression and Female Agency

One researcher that has been especially important for shifting perceptions about female birds, and female animals in general, is ornithologist Patricia Adair Gowaty. I interviewed her in her home in Topanga, a mountainous area just northwest of Los Angeles. Gowaty is a Distinguished Professor Emerita at University of California Los Angeles, with a PhD from Clemson University, South Carolina (1980). Her main study species is the eastern bluebird, *Sialia sialis,* and she has conducted pioneering research in several areas of sexual selection, both theoretically and empirically, on polyandry, female aggression and flexible mate choice.

Gowaty tells me that during her time studying at a women's college, she had been involved in discussions about racism, the Vietnam war and civil rights, but did not yet pay attention to sexism and feminism. She says "I was pretty deaf then to paying attention to female animals". Gowaty recalls that it was when she worked with science education at Bronx Zoo, in New York, and joined a feminist conscious-raising group in 1969 that she simultaneously emerged as a scientist and a feminist. At the zoo, she was surrounded by powerful and influential women biologists like Kathrine Ralls[68] and Jocelyn Crane.[69] They talked about women

[67] Keller, 2004: 8.

[68] Ralls published a pioneering paper in 1976 about mammals in which females are larger than males, until then the phenomenon had been overlooked, and males generally assumed to be the larger sex.

[69] A senior researcher and expert on fiddler crabs.

in science and female control of reproduction—"part of the discussion about female animals then was powered by the fact that many of us were very well aware of what was going on in the lives of women". Current public debates raged about the abortion court case Roe vs Wade, which ended with the US Supreme court (1973) deciding that abortion is part of women's constitutionally protected right to privacy.[70] Gowaty says, "I really am interested in females, in female biology. That's no surprise, I'm a female—I'd like to know more about that." Thus, I interpret that she connects her interest in females to her own situatedness as a female. Similar to feminist science philosopher Evelyn Fox Keller's description of geneticist Barbara McClintock's deep and intimate understanding of corn, as an absorption in and identification with her material,[71] with a *feeling for the organism,* Gowaty relates to females of different organisms and their respective constraints. In terms of harassment from males, she considers female mammals to be most exposed, while birds can often just fly away. Similarly, comparing the constraints of females among different species of flies, she exclaims "If you're going to be a fly, you ought to be a Hawaiian *Drosophila*!" She explains that because they mate at one place and lay eggs at another, Hawaiian *Drosophila* females are at leisure to mate with whomever they want, without being harassed. In contrast, their common relative the fruit fly *Drosophila melanogaster,* mate and lay eggs at the same place, so that males harass females while they oviposit and females do not have the same freedom. This feminist concern and gender-reflective approach of Gowaty's has led her to theorize and test hypotheses about social and ecological constraints on females and males, and female resistance to control by males of their sexuality and reproduction.

## Reporting Polyandry in a Socially Monogamous Bird

In the history of turning perceptions about females, one important part is the reception of new ideas and findings about females in the scientific community. In 1984, Patricia Gowaty and her collaborator Alvan Karlin

[70] A decision that has been challenged repeatedly as several states have enacted restrictions on abortions, and as of 24 June 2022 has been overruled by the Supreme Court.

[71] Keller, 1983: xiv.

published a controversial study, namely the first report of female multiple mating in a socially monogamous bird, the eastern bluebird.[72] After hearing about the polyandrous females, Stephen Emlen, a world authority of the social behavior of birds, exclaimed: "Those females were raped!"[73]—in accordance with prevailing assumptions of passive females. Gowaty's interpretation differed, according to her, the nature of passerine bird copulations—that they (as most birds) lack a penis to transfer sperm, instead sperm is transferred between cloacas (the combined opening of the digestive, urinary and reproductive tract), and copulations occur very quickly and females can fly away to avoid them—means that it is virtually impossible to force copulate a female passerine.[74]

Gowaty recalls that many colleagues were skeptical to the new findings of female multiple mating and assumed that it would not happen in their study species, as there was little behavioral evidence. As mentioned above, the prevailing theoretical assumptions were that females were monogamous—because 90% of birds are socially monogamous it was generally expected that birds were genetically monogamous—especially the females, who according to theory had nothing to gain from multiple mating. Thus, as described above, initially the scientific community generally interpreted female multiple mating in birds as male-induced, and Gowaty's interpretation of females as active was in the minority.

In 1991, skepticism was yet again the response when Gowaty and coworkers presented findings that bluebird females actively seek extra-pair copulations.[75]

## Surrounded by Parentage Detectives

There are many different influences forming researchers' situated knowledges, for example, their institutional situation, which questions were

[72] Gowaty & Karlin, 1984.

[73] There has been a long controversy over anthropomorphic terminology such as "rape" in animal behavior (for example Gowaty, 1982). "Cohersion" or "forced copulation" have been suggested as alternative terms.

[74] Patricia Gowaty interview 2017.

[75] Gowaty & Bridges, 1991.

currently pursued by their local scientific community, influences from other disciplines and the available technologies. At the time when Gowaty was doing the parentage study on the eastern bluebirds, she had many colleagues that worked on the same question—who the parents were among different animals. One of them, the well-known evolutionary geneticist John Avise, investigated parentage in many different species, for example, in fishes, mammals and birds, through examining allozymes (variations in enzymes coded at the same locus in the DNA, the same method that Gowaty and Karlin used). Moreover, population geneticist Wyatt Anderson, who was Gowaty's mentor and colleague, had published a paper on female multiple mating in fruit flies, *Drosophila pseudoobscura,* already in 1974, using chromosome inversions as a method to trace paternity[76] (i.e., identifying rearranged segments of chromosomes in certain individuals). Technical innovations, such as molecular techniques for determining parentage were important for the development of research in sexual selection, they opened up for investigating questions that had not been possible before. As described in Chap. 3, in population genetics, female multiple mating in fruit flies was well known in the late 1960s. I contacted Wyatt Anderson to ask about his eventual experiences of resistance in the 1970s, but he could not recall any resistance to his paper on female multiple mating in wild fruit flies,[77] in contrast to the skepticism Gowaty met to her findings of female multiple mating in birds in the 1980s. Anderson confirms that the lack of resistance in population genetics was not surprising given renowned population geneticist Theodosius Dobzhansky's earlier laboratory studies on female multiple mating. Here is also an example of where knowledge about female multiple mating in one field, population genetics, has transferred to the animal behavior and sexual selection studies.

Thus, Gowaty was in a research environment that both had the interest for these parentage questions, including a population geneticist that had demonstrated female multiple mating, and in which she could find a collaborator that had the skills for conducting the laboratory analyses. Thus,

---

[76] Anderson, 1974.

[77] Wyatt Anderson, Personal Communication 2019.

Gowaty's institutional situation provided her with both inspiration and practical possibilities to pursue questions about parentage.

## Aggressive Females

As described above, assumptions about females as passive and peaceful have hindered knowledge production about females. These prevailing assumptions were also affecting the response toward pioneering discoveries of female aggression. In one of Gowaty's early experimental studies, she demonstrated fierce and sometimes deadly female-female aggression in eastern bluebirds, which she documented in photographs. This result went against the grain of the common assumption of females as non-violent and a renowned ornithologist reacted with incredulity, "I wouldn't have believed it, if I hadn't seen the pictures".[78] Thus, pioneering findings about females were repetedly received with skepticism in the scientific community.

Furthermore, researchers' situatedness as humans, with differences in reproductive biology compared to birds, may have influenced assumptions about female birds. In Gowaty's report, where she connects the female-female aggression with defense against other females laying eggs in their nests (nest parasitism), she notes: "Curiously, ornithologists have only recently started to study nest parasitism within one species and to ponder its consequence—uncertain maternity. Could it be that we have ignored the potentially strong selective force of alternative female reproductive strategies because we are mammals, animals for which only paternity is questionable?"[79] In accordance with my analysis of situated knowledges, this example suggests one more way in which researchers' understanding of the world is influenced by their situatedness as humans—because the researchers are mammals, for whom which female has given birth to which young is certain, the idea that other organisms don't have the same maternity certainty may have been delayed and contributed to the ignorance of female aggression when at risk for nest parasitism. This

[78] Gowaty, 2010:324.

[79] Gowaty, 1985a:12.

situated ignorance may have contributed to perpetuate the ignorance about how females compete among each other in birds and other egg-laying organisms.

## Feminist Influences and Interventions

Gowaty's work shows how feminist interventions have been important for shifting perceptions about females in sexual selection. She has frequently used current feminist ideas to generate testable scientific hypotheses.[80] Moreover, she was early on engaged in debates about use of terminology in the field, arguing against the use of sexually evocative terms, such as "rape".[81] In 1994, she arranged the *Feminism and Evolution Symposium* at the University of Georgia, supported by the Society for the Study of Evolution as well as the University of Georgia. The conference provided an opportunity for many scholars to jointly discuss gender bias in the field, although Gowaty recalls that some participants were hostile toward each other.[82] She edited the subsequent book *Feminism and Evolutionary Biology—Boundaries, Intersections and Frontiers,* which did not get much attention according to Gowaty, although Sarah Blaffer Hrdy describes the conference as influential.[83] Yet, the precursor of the conference was a small gathering that had taken place in 1993 with thirteen feminist evolutionary biologists who discussed their experiences, evolutionary biology and their commitments to end sexist oppressions.[84] Both before and after the conference, this informal network of scientists challenged prevailing perceptions about females, sometimes publishing collectively,[85] and each in their own sub-field, drawing new cultural maps of what females were and could do. In parallel with Nancy Tuana's analysis of the Women's Health Movement, these feminist biologists exert epistemological

[80] Gowaty, 2010: 333.
[81] Gowaty, 1982.
[82] Gowaty, 2010:333.
[83] Hrdy, 1999: xviii.
[84] Gowaty, 1997b: xv.
[85] For example Fausto-Sterling et al., 1997.

resistance through undermining the production of ignorance about females and by creating liberatory knowledges.[86]

## Pushing Boundaries

Due to her pioneering work and the resistance it met, Gowaty had a hard time getting a permanent position in the beginning of her career; it took twelve and a half years after her PhD before she got a tenure-track position at the University of Georgia.[87] The position she got was as Associate Professor in Zoology and Women's studies at the University of Georgia, a position founded by a professor in Ecology that decided that Ecology was to be integrated with each and every other discipline. Here, her unusual combination of expertise served as an opportunity for institutional establishment. She was soon promoted to Full Professor in Zoology and pursued a successful career and later received a distinguished professorship at UCLA.

Gowaty pushes the boundaries of what sexual selection means by challenging assumptions, arguing for reinterpretations and paying attention to females. She has been pivotal in challenging the Darwin-Bateman paradigm with her critique[88] and repetition of Bateman's experiment,[89] and together with her husband and renowned ecologist Stephen Hubbell, she has developed gender-neutral models of mate choice,[90] alternatives to the prevailing paradigm. I interpret Gowaty's strategy as primarily in line with what science philosopher Sandra Harding categorizes as feminist empiricists,[91] namely to argue for increased awareness of gender bias in combination with rigorous scientific empiricist methods, formulation of testable hypotheses and "blind" observations[92] (i.e., the observer is unaware of the theoretical prediction). Harding describes feminist

[86] Tuana, 2006: 2.

[87] Gowaty, 2010:330.

[88] Snyder & Gowaty, 2007.

[89] Gowaty et al., 2012.

[90] Gowaty & Hubbell, 2005, 2009.

[91] Harding, 2006: 237f.

[92] Gowaty, 2003:917.

empiricism as the insistence that the "bad science" that emerges from gender bias in assumptions and interpretations is expected to be eliminated by more rigorously following existing empiricist methods and norms in science. However, Gowaty also discusses the role of feminism for changing science (by including women, by spreading feminist consciousness, by feminist ideas inspiring formulation of testable hypotheses and by providing better controls for experiments) and the benefits of collaborating with researchers that do not share the same political commitments and assumptions.[93] Thus, Gowaty also includes aspects of what Harding categorizes as philosophical feminist empiricism, in line with science philosophers Helen Longino's and Lynn Hankinson-Nelson's different epistemologies, namely that social values are inescapable in the content of science and may sometimes have positive influences.[94]

Gowaty's situatedness, as a feminist and scientist, with her institutional position which provided collegial interest and skills in parentage studies, enabled her to develop gender-reflective approaches and challenge prevailing sexual selection theory and research. Furthermore, her study species the eastern bluebird, with its variable mating and parental care patterns, provided the basis for theorizing about dynamics between female and male strategies as dialectics[95] (see below in *the Influence of Study Species studyspecies*). Gowaty says that her aim is not merely to bring forward female perspectives, but to make an egalitarian effort of understanding constraints for both females and males. She says what is needed is theories that account for variation among individuals beyond sex.[96] She conceives of females as active, soliciting matings, influencing which sperm fertilize their eggs, variable among themselves, sometimes aggressive among themselves and dominant to females as well as males, and assumes that some or most females are taking care of themselves.[97] As we have seen, even though feminist biologists perceive females as active and challenge prevailing perceptions in the wider scientific community and thereby are important agents for change—such interventions are

[93] Gowaty, 2003: 917.
[94] Harding, 2006: 237f.
[95] Gowaty, 1997c.
[96] Gowaty, 1997a, Gowaty & Hubbell, 2005, 2009.
[97] Gowaty, 1997a:96.

often met with skepticism or ignored. Thus, through the negotiation of scientific knowledge, such interventions have in many cases changed scientific thinking and the content of science.[98]

## Simon Griffith—From Extra-Pair Paternity to Pair-Bonds

Ornithologist Simon Griffith is a Professor at Macquarie University, Sydney, where I met him for an interview in his office[99] (Fig. 4.1). His research and perspectives are interesting to the history of the female turn because his own work has developed from classical sexual selection studies of ornamentation and extra-pair paternity to challenging perceptions of females and the relationship between the sexes. His account also illustrates how the researcher's study species influence their situated knowledges.

### Studying Sexual Selection in European Model Species

Griffith started his research focusing on extra-pair paternity in the European model species of birds—house sparrows (*Passer domesticus*), blue tits (*Cyanistes caeruleus*), and collared flycatchers (*Ficedula albicollis*). He gained his PhD in 1998, studying house sparrows in an island population that showed no extra-pair paternity. Then, he spent four years measuring blue tit ornamentation in males and females during his postdoc in Oxford, UK. Blue tit is one of the most studied species when it comes to sexually selected ornamentation. He and his colleagues collected a huge dataset, but the results showed that "basically nothing [was] going on at all" and they found it very hard to publish this against-the-grain conclusion. Thus, testifies Griffith, studies that confirm the current prevailing understanding of sexual selection, such as studies showing sexual selection in male blue tit ornamentation, are much easier to publish than

[98] Schiebinger, 1999:181ff.

[99] Simon Griffith interview 2018.

**Fig. 4.1** Ornithologist Simon Griffith in his office at Macquarie University, Sydney, Australia

studies that argue for the oppositional interpretation that blue tit ornamentation is not sexually selected.

In contrast, Griffith's next study species, the collared flycatchers in Sweden, showed very strong sexual selection. Females mated with multiple males, and the highest level of polyandry was found in the females paired with males from the sister species, the pied flycatcher (*Ficedula hypoleuca*). Griffith understands this strong selection in the collared flycatcher not as an effect of competition between males within that species (sexual selection), but resulting from the strong selection on collared flycatcher females to always mate with somebody different from their partner in order for the offspring to be fertilized by a male of their own species.[100] Hence, Griffith interprets this case as female multiple mating being driven by the presence of the pied flycatcher. Thus, according to Griffith's interpretation "not much is going on" in sexual selection except in extreme cases. This viewpoint is somewhat apart from the mainstream sexual selection community in which sexual selection is seen as a generally important evolutionary force[101] and that ornaments and birdsong are often sexually selected.[102]

## Studying Australian Birds

Griffith's move to Australia further shifted his perspectives on sexual selection, and his understanding of the relationship between males and females and developed his gender-reflective approaches. Here is another example that the study species, and their geographical location, influence the researcher's situated knowledges. Griffith explains that most Australian birds are long-lived, sing a lot of duets (i.e., females and males take turns singing, in contrast to traditional assumptions about male birdsong as an adaptation for displaying toward females and marking territoriality against other males) and spend a lot of time preening each other, which has led him to investigate the role of the pair-bond. He is now inclined to think that choosing a partner due to a certain ornamental trait is not that important (sexual selection for ornaments), instead the collaboration around reproduction is more important.

One of his current study species is the zebrafinch (*Taeniopygia guttata*), in which the individuals in a couple have continuous song conversation with each other over years, thus, he says, singing cannot be only for the sake of mate choice. Furthermore, females' bill color vary just as much as males'. Former studies were all about male bill color—thus the research

---

[100] Collared flycatcher females mated with pied flycatcher males have higher levels of extra-pair nestlings fathered by males of their own species (Veen et al., 2001).

[101] See for example Lindsay et al., 2019.

[102] See for example Andersson & Iwasa, 1996.

community's focus on male ornamentation led to ignorance of female ornamentation. These studies of bill color also exemplify male precedence in sexual selection research, which was also the case for other ornamental traits[103] (Chap. 3).

Griffith says his motivation is interest in answering questions in a neglected area, setting "wrongs to right" and "add some important insight". He is not interested in doing another "stamp-collecting" confirmatory study of extra-pair paternity that doesn't lead to new insights. Thus, his motivation shows similarities with Patricia Gowaty's—though he does not mention being motivated by a feminist incentive as Gowaty—he too is interested in making scientific progress.

## Experiences of Shifting Perceptions

Griffith agrees that a "polyandry revolution" has taken place in terms of molecular techniques allowing revolutionary progress in measuring parentage. But he doesn't think that there have been any major tipping points in the consensus opinion over time, and he's also "not clear that everyone is in agreement now about where things are at". He says that some people have a constant worldview that doesn't change. According to Griffith, "things have gone up and down" with regard to perceptions about females in bird research. Rather than a tipping point, it was an incremental amassing of evidence from the late 1980s until the new consensus was reached that extra-pair paternity is quite widespread, even across monogamous species of birds. Thus, there has been a shift from perceptions of females as principally monogamous to most species demonstrating some degree of female multiple mating. Although, as Griffith points out, this shift does not mean that the categorization of mating systems (the pattern of sexual interactions between individuals that is typical of a species[104]) has changed much, species that were formerly categorized as monogamous are still largely categorized as monogamous, even though we now know that they have a few percent extra-pair offspring. According to Griffith, the more interesting question is why

[103] Ah-King 2022.

[104] Leonard, 2019: 584.

extra-pair paternity isn't more frequent and why so many birds actually have very good pair bonds and very low levels of extra-pair paternity.

However, even if categorizations of species as monogamous or polygamous have not changed significantly, there has been a shift in that almost all females are now perceived as potentially mating with multiple males. I interpret this shift in perceptions about females as one aspect of the female turn in bird research.

## Controversy over Interpretations of Female Agency in Extra-Pair Copulations

In 2002, Griffith and coworkers published an influential review, demonstrating the extent and variability of extra-pair paternity among birds. But he says that most of the reviewed studies were focused on males, though implicitly that empirical data provided information about what females are doing. There are very few studies that have investigated actual copulations, because that is very hard to do in birds, and there seems to have been a lack of interest. Griffith thinks that many researchers are comfortable with the idea that males are driving polyandry and that females are participating in a quite passive way. The debate about whether to perceive females as active or not is still going on.

Griffith gives an example of the lingering perceptions of passive females in a well-cited review about causes of "female infidelity" in socially monogamous birds by evolutionary geneticist and animal ecologist Göran Arnqvist and theoretical population geneticist Mark Kirkpatrick (2005)[105] (infidelity is another anthropomorphic term which implicitly indicates that there is a contract of exclusive sexual monogamy between partners). The authors argue that the best interpretation of the reviewed empirical studies is that females do not gain indirect benefits from extra-pair matings and that males are selected to increase their reproduction through extra-pair fertilizations. In other words, they interpret extra-pair copulations as an outcome of sexually antagonistic coevolution (sexual

[105] Arnqvist & Kirkpatrick, 2005.

conflict),[106] in which these extra-pair matings are not adaptive for females. However, Griffith does not agree with this interpretation, because many alternative hypotheses are ignored.[107] For example, the reviewed data cannot refute the hypothesis that females are mating with multiple males to assure fertility (which would be adaptive for females). Even though the fertility assurance hypothesis is hard to test empirically, it might be central for explaining low levels of extra-pair copulations. In the production of ignorance about females, this is one example of alternative hypotheses being overlooked. The focus on only empirical evidence and neglecting theoretical possibilities leads to lingering perceptions of passive females. As I will come back to in Chap. 6, the framework of sexual conflict has been, both theoretically and empirically, largely focused sexual conflict that benefits males to the detriment of females.

Furthermore, there is a well-cited follow-up study to the review mentioned above that investigates sexually antagonistic co-evolution on the genetic level, claiming support for the idea that "genes-for-male-promiscuity-causes-female-promiscuity".[108] Wolfgang Forstmeier et al. (2011) used zebra finches and found a genetic correlation between male attempts of extra-pair copulations and female acceptance of copulations.[109] Griffith describes it as a very important paper with "very good sample size, huge data set and they got a nice pedigree and they demonstrate there's real genetic correlation between those". However, he doesn't agree with the authors' interpretation that males are dragging along females toward promiscuity. Griffith says there is evidence of a genetic correlation, but the *causation* is not clear, it could also be the females dragging the males along. Female Zebra finches in captivity have for generations been selected by breeders and researchers to mate quickly. Again, an alternative hypothesis that suggests an active role of females is overlooked. Griffith is challenging prevailing perceptions of females and males in sexual selection—thus the debate continues.

---

[106] See also Chap. 6 on the development of sexual conflict and the controversy between cryptic female choice and sexually antagonistic coevolution.

[107] Griffith, 2007.

[108] Simon Griffith interview 2018.

[109] Forstmeier et al., 2011.

## Cryptic Female Choice in Birds

In the history of sexual selection, sperm competition, the suggestion of the male-focused hypothesis of post-copulatory sexual selection precedes the proposition of the equivalent female-focused hypothesis of cryptic female choice (females influencing which sperm fertilizes their eggs during or after copulation; see Chap. 6). Griffith has conducted experiments on cryptic female choice in birds. He asserts that it is extremely hard to prove that cryptic female choice occurs in birds, very few experiments have tried to test it and when it is done, he knows by own experience, it is very hard to publish.

Although a few ornithologists are engaged in testing cryptic female choice hypotheses about behavior and physiology, it is extremely difficult to find good evidence in birds, probably much more challenging than in many other organisms, for example insects, resonates Griffith. He doesn't think that most ornithologists think about females as active—most studies are done from an androcentric perspective and females are still overlooked. He argues that it is much easier to study the male part in sexual selection—compare testis size between species or investigate sperm morphology or performance—than following marked females in the field for several years or study what happens inside their internal organs. Despite new technologies, we cannot expect the field to shift because of this bias in the ease by which studies are made. Hence, he argues, there will continue to be a bias toward male-focused compared to female-focused studies of sexual selection in birds.

Hence, in the history of the female turn in ornithology, Griffith's account testifies the continued male bias and production of ignorance about females in studies of sexual selection in birds, the ignorance of alternative hypotheses and interpretations, and the resistance to publish studies that oppose the currently prevailing understanding of sexual selection. His situated knowledges are formed by his empirical sexual selection studies on classical European model species, and later studies of Australian birds—which motivated his challenges of prevailing perceptions about females and relationships between the sexes. Thus, he has contributed to knowledge about active females, and become involved in debates about ignorance of females in the field.

## The Influence of Study Species

In the history of the female turn, I found that the study species is one important influence in forming the researchers' situated knowledges. For example, ornithologist Richard Wagner studied razorbills, *Alca torda*, and compared them with their relatives the common guillemots, *Uria aalge*—in the former species females have long tails that enable them to physically hinder unwanted copulations, while guillemot females have shorter tails, breed in high-density colonies and cannot hinder forced copulations.[110] Hence, Wagner discussed these differences in constraints on females in the two species, and concluded that female razorbills can actively control extra-pair copulations.[111] In contrast, Tim Birkhead and coworkers who studied common guillemots, watched males vigorously mate guarding and females most often attempting to resist extra-pair copulations, they interpreted extra-pair copulations as a male strategy, and the occasional female acceptance as a female cost-reducing strategy or a sign of early new pair-formation[112]—in line with prevailing assumptions of monogamous females and promiscuous males. Thus, the study species is an important influence in forming the researchers' situated knowledges, leading to divergent understandings of sex differences and sexual selection. (Although an investigator with a feminist strategy might have focused more on those fewer instances when female guillemots were initiators to extra-pair copulations, a later study showed that these female-initiated copulations more often led to cloacal contact and therefore were more likely to produce offspring.[113]) The importance of the study species in my analysis of the female turn is in line with science analyst Bruno Latour's emphasis on the agency of

[110] Wagner, 1991:163.

[111] Wagner, 1991.

[112] Birkhead et al., 1985: 617.

[113] Hatchwell, 1988.

**Fig. 4.2** (A) A pair of eastern bluebirds, *Sialia sialis*, photo by Sandysphotos Creative CommonsAttribution 2.0 Genericlicense. (B) A pair of Australian zebra finches, *Taeniopygia guttata castanotis*, photo by Lip Kee Yap Creative CommonsAttribution-Share Alike 2.0 Genericlicense

non-human actors[114] in the scientific process, as well as with Donna Haraway's description of the world as a "witty agent"[115]. The study animals are not mere passive objects, but have agency in the knowledge-making process. However, since there were early studies describing active females (e.g., Harry Hann, 1937),[116] that were later re-interpreted as possibly "apparent rather than real"[117] (an example of ignorance production), the researcher's interpretation is also influenced by other aspects of their "specific *ways* of seeing",[118] such as the theoretical focus on males in sexual selection or alternatively, feminist critique and gender-reflective approaches (Fig. 4.2).

The geographical location of study species is also an important influence on researchers' situated knowledges. Simon Griffith's account provided one example, his work on long-lived Australian birds, such as zebra finches, affected his understanding of interactions among the sexes and the importance of the pair-bond. In a similar line, evolutionary biologist Jordan Price notes that to Charles Darwin, male birdsong was a textbook

[114] Latour, 2005.

[115] Haraway, 1988: 593.

[116] Hann, 1937.

[117] McKinney et al., 1984: 538.

[118] Ibid: 583.

example of sexual selection, but "if Darwin had grown up in the tropics or if he was Australian, he might have asked why the female birds he saw on his travels in northern climates *didn't* sing instead of asking why males have elaborate song."[119] Because in the Northern Hemisphere, where most bird research has been conducted, it is mainly the male songbirds that sing. In contrast, a recent study by Karan Odom and co-workers (2014),[120] revealed that female song in songbirds[121] is widespread, especially in the tropics and in Australia, and has been much overlooked. Traditionally in birdsong research, female song was dismissed as atypical or the effect of hormonal disruption.[122] Again, as in the research on visual ornaments, the primary theoretical focus on males in sexual selection produced ignorance and delayed knowledge about females' equivalent characteristics, hence this is another example of male precedence. Furthermore, Odom and colleagues' reconstruction shows that both males and females singing is ancestral to songbirds. Thereby they challenge the traditional interpretation of birdsong as primarily sexually selected in males, instead they emphasize the question why female song has been lost in certain lineages and whether social[123] or natural selection can explain the evolution of birdsong in both sexes. In this case, the production of ignorance about females reflects both androcentrism and a geographical bias. Furthermore, as mentioned in Chap. 3, especially women ornithologists have contributed to this discovery.[124] Again, we can see how the different aspects of the researchers' "specific *ways* of seeing" affect their understanding of females.

Furthermore, the study species also influences the researchers' understanding of relations between the sexes. Ornithologist Marion Petrie discussed this issue and brought up two contrasting examples.[125] Petrie

---

119 Borgmann, 2019.

120 Odom et al., 2014.

121 Songbirds are species within the suborder Passeri (oscine passerines), including, for example, crows and lyrebirds.

122 Odom et al., 2014: 2.

123 Social selection is selection resulting from social interactions, which includes sexual selection (West-Eberhard, 1983), see also chap. 7.

124 Haines et al., 2020.

125 Marion Petrie interview 2018.

herself studied peafowl (*Pavo cristatus*), that is peacocks and peahen, in which males are spectacularly ornamented and contribute sperm, but no parental care, and thus there is not much interaction between the sexes apart from mating. Then, she compares with Patricia Gowaty's perspective influenced by studying eastern bluebirds:

> "I think her perspective comes from working on the paired organism, so that you've got [a] male-female dynamic where they're sharing care duties, whereas for me… I'm left with wondering what the point of a male peacock is, they don't do anything, they're just quintessentially the useless individuals. So, I don't have a female resistance to male control dynamic in my own organism."[126]

Thus, Petrie contrasts her own understanding of male-female (nearly non-existing) relations to Gowaty's idea of dynamic male-female evolutionary dialectics.[127] In Gowaty's study species the eastern bluebird, females and males bring up offspring together, thus there is a lot of interaction between the sexes. Hence, Gowaty has developed a view on the sexes as dynamically interacting and co-evolving. This difference in knowledge does not mean that one is more right than the other; it means that Petrie and Gowaty through their chosen study species have gained different partial knowledges about interactions between the sexes in these specific bird species. To understand the full range of interaction between sexes we need to study the full diversity of species.

This analysis of the female turn shows the important influence of the study species—along with feminist insights, gender-reflective approaches, theoretical framework, institution and methodology—for forming the researchers' different understandings of sex and sexual selection.

---

[126] Marion Petrie interview 2018.

[127] For example Gowaty, 1997c.

## Chapter Conclusions

As the paradigmatic animals to study in sexual selection, birds have received extensive research attention. The starting point was the assumption that most birds—especially females—were monogamous. The application of new DNA fingerprinting technology to analyze bird parentage revealed much more widespread female multiple mating than predicted. Initially, these findings were largely interpreted as caused by male strategies and females being passive. Although there were some early empirical studies describing female birds as actively pursuing copulations with multiple males, such studies became more frequent in the 1980–90s, authored by women as well as men, contributing to shifting perceptions about females. However, women feminist ornithologists especially formulated an explicit gender-reflective approach of questioning the prevailing assumptions of passive females. Thus, their research undermined the production of ignorance about females.

In bird research, as well as in primatology and in sexual selection research in general, ignorance has been produced through the recurring pattern of male precedence—that research starts with male-centered investigations or explanations and later includes the female-centered equivalents—in ornament evolution, in interpretations of female multiple mating, in research about dominance, bill color and birdsong. Furthermore, ignorance has been produced through the loss of acquired knowledge—when researchers casted doubt on early interpretations of female active participation in multiple mating; by overlooking alternative hypotheses—such as the fertility assurance hypothesis, and by not citing—for example, studies on female breeding dominance.

The community of bird researchers includes widely different perceptions about females, thus debates about whether to interpret females as passive or active continue. I interpret these differences among researchers as due to their situated knowledges and ignorance—formed by their sexed/gendered experiences, their theoretical perspectives, their institutional position, their methodological opportunities, their geographical location, their feminist perspectives, and their knowledge about specific study species. Patricia Gowaty has used her feminist insights to generate hypotheses and theories about female and male strategies in sexual selection, and her primary study species, the eastern bluebird, seems to have been important for her theorization of female and male sexual selection strategies as a dynamic "dialectic". Also for Simon Griffith, his study species, European as well as Australian birds, have influenced his interpretations of female agency and the significance of the pair-bond. Several additional examples indicate that the study species forms an important aspect of a researcher's situated knowledges.

Due to their specific situated knowledges, certain researchers have challenged prevailing assumptions in sexual selection, confronted ignorance about females and constructed new cultural maps of females and their agency. Yet, production of ignorance about females—due to male precedence, as well as the general androcentrism in sexual selection and also empirical difficulties in investigating sexual selection in females—continues in sexual selection research on birds.

## References

Ah-King, M. (2022). The history of sexual selection research provides insights as to why females are still understudied. *Nature Communications 13*(1) 6976. http://doi.org.10.1038/s41467-022-34770-z

Anderson, W. W. (1974). Frequent multiple insemination in a natural population of *Drosophila pseudoobscura*. *The American Naturalist, 108*(963), 709–771.

Andersson, M., & Iwasa, Y. (1996). Sexual selection. *Trends in Ecology and Evolution, 11*(2), 53–58.

Arnqvist, G., & Kirkpatrick, M. (2005). The evolution of infidelity in socially monogamous passerines: The strength of direct and indirect selection on extrapair copulation behavior in females. *The American Naturalist, 165*, S26–S37.

Barrow, M. V. (1998). *A passion for birds: American ornithology after Audubon*. Princeton University Press.

Birkhead, T. R. (1987). Sperm competition in birds. *Trends in Ecology & Evolution, 2*, 268–272.

Birkhead, T. R. (2000). *Promiscuity: An evolutionary history of sperm competition and sexual conflict*. Faber.

Birkhead, T., & Montgomerie, R. (2020). Three decades of sperm competition in birds. *Philosophical Transactions of the Royal Society of London B, 375*, 20200208. https://doi.org/10.1098/rstb.2020.0208

Birkhead, T. R., Johnson, S. D., & Nettleship, D. N. (1985). Extra-pair matings and mate guarding in the common murre Uria aalge. *Animal Behaviour, 33*, 608–619.

Borgmann, Kathy. (2019). The forgotten female: How a generation of women scientists changed our view of evolution. *The living bird magazine* June 17, 2019.

Bray, O. E., Kennelly, J. J., & Guarino, J. L. (1975). Fertility of eggs produced on territories of vasectomized red-winged blackbirds. *Wilson Bulletin, 87*, 187–195.

Burke, T., & Bruford, M. W. (1987). DNA fingerprinting in birds. *Nature, 327*, 149–152.

Burley, N. T., & Parker, P. G. (1998). Emerging themes and questions in the study of avian reproductive tactic. In P. G. Parker & N. T. Burley (Eds.), *Avian reproductive tactics: Female and male perspectives. Ornithological monographs no. 49*. The American Ornithologists' Union.

Double, M., & Cockburn, A. (2000). Pre-dawn infidelity: Females control extra-pair mating in superb fairy-wrens. *Proceedings of the Royal Society: Biological Sciences, 267*(1442), 465–470.

Fausto-Sterling, A., Gowaty, P. A., & Zuk, M. (1997). Evolutionary psychology and Darwinian feminism. *Feminist Studies, 23*(2), 403–417.

Fitch, M. A., & Shugart, G. (1984). Requirements for a mixed reproductive strategy in avian species. *American Naturalist, 124*(1), 116–126.

Forstmeier, W., Martin, K., Bolund, E., Schielzeth, H., & Kempenaers, B. (2011). Female extrapair mating behavior can evolve via indirect selection on males. *Proceedings of the National Academy of Science USA, 108*(26), 10608–10613.

Gowaty, P. A. (1982). Sexual terms in sociobiology—Emotionally evocative and paradoxically, jargon. *Animal Behaviour, 30*, 630–631.

Gowaty, P. A. (1985a). Bluebird belligerence. *Natural History, 6*(85), 8–12.

Gowaty, P. A. (1985b). Multiple parentage and apparent monogamy in birds. In P. A. Gowaty & D. W. Mock (Eds.), *Avian monogamy* (pp. 11–21). Ornithological Monographs No. 37.

Gowaty, P. A. (1997a). Principles of females' perspectives in avian Behavioral ecology. *Journal of Avian Biology, 28*(2), 95–102.

Gowaty, P. A. (Ed.). (1997b). *Feminism and evolutionary biology, boundaries, intersections and frontiers*. Chapman and Hall.

Gowaty, P. A. (1997c). Sexual dialectics, sexual selection and variation in reproductive behavior. In P. A. Gowaty (Ed.), *Feminism and evolutionary biology, boundaries, intersections and frontiers* (pp. 351–384). Chapman and Hall.

Gowaty, P. A. (2003). Sexual natures: How feminism changed evolutionary biology. *Signs, 28*(3), 901–921.
Gowaty, P. A. (2010). Watcher: The development of an evolutionary biologist. In L. C. Drickamer & D. A. Dewsbury (Eds.), *Leaders in animal behavior, second generation* (pp. 309–342). Cambridge University Press.
Gowaty, P. A., & Bridges, W. C. (1991). Behavioral, demographic, and environmental correlates of uncertain parentage in eastern bluebirds. *Behavioral Ecology, 2*, 339–350.
Gowaty, P. A., & Hubbell, S. P. (2005). Chance, time allocation, and the evolution of adaptively flexible sex role behavior. *Integrative and Comparative Biology, 45*, 931–944.
Gowaty, P. A., & Hubbell, S. P. (2009). Reproductive decisions under ecological constraints: It's about time. *Proceedings of the National Academy of Science USA, 106*(Suppl. 1), 10017–10024.
Gowaty, P. A., & Karlin, A. A. (1984). Multiple maternity and paternity in single broods of apparently monogamous eastern bluebirds (Sialia sialis). *Behavioural Ecology and Sociobiology, 15*(2), 91–95.
Gowaty, P. A., Kim, Y.-K., & Anderson, W. W. (2012). No evidence of sexual selection in a repetition of Bateman's classic study of drosophila melanogaster. *Proceedings of the National Academy of Science USA, 109*(29), 11740–11745.
Gray, Elizabeth M. (1994). The ecological and evolutionary significance of extra-pair copulations in the Red-winged Blackbird (*Agelaius phoeniceus*). Ph.D. dissertation, University of Washington, Seattle.
Gray, E. M. (1997). Female red-winged blackbirds accrue material benefits from copulating with extra-pair males. *Animal Behaviour, 53*, 625–639.
Gray, E. M. (1998). Intraspecific variation in extra-pair behavior of red-winged blackbirds (Agelaius phoeniceus). In P. G. Parker & N. T. Burley (Eds.), *Avian reproductive tactics: Female and male perspectives. Ornithological monographs no. 49. Washington DC.* The American Ornithologists' Union.
Griffith, S. C. (2007). The evolution of infidelity in socially monogamous passerines: Neglected components of direct and indirect selection. *American Naturalist, 169*(2), 274–281.
Griffith, S. C., Owens, I. P. F., & Thuman, K. A. (2002). Extra pair paternity in birds: A review of interspecific variation and adaptive function. *Molecular Ecology, 11*, 2195–2212.
Haines, C. D., Rose, E. M., Odom, K. J., & Omland, K. E. (2020). The role of diversity in science: A case study of women advancing female birdsong research. *Animal Behaviour, 168*, 19–24. https://doi.org/10.1016/j.anbehav.2020.07.021

Hann, H. W. (1937). Life history of the oven-bird in southern Michigan. *Wilson Bulletin, 49*, 145–237.

Haraway. (1986). Primatology is politics by other means. In R. Bleier (Ed.), *Feminist approaches to science* (pp. 77–118). Pergamon Press.

Haraway, D. J. (1988). Situated knowledges: The science question in feminism and the privilege of partial perspectives. *Feminist Studies, 14*(3), 575–599.

Harding, S. (2006). Rethinking standpoint epistemology: What is 'strong objectivity'? In E. F. Keller & H. E. Longino (Eds.), *Feminism and science* (pp. 235–248). Oxford University Press.

Hatchwell, B. J. (1988). Intraspecific variation in extra-pair copulation and mate defence in common guillemots Uria Aalge. *Behaviour, 107*(3–4), 157–185.

Hrdy, Sarah Blaffer. (1999). [originally published 1981]. The woman that never evolved. Harvard University Press.

Keller, E. F. (1983). *A feeling for the organism: The life and work of Barbara McClintock*. W. H. Freeman.

Keller, E. F. (2004). What impact, if any, has feminism had on science? *Journal of Biosciences, 29*(1), 7–13.

Kempenaers, B., Verheyen, G. R., Vanderbroeck, M., Burke, T., & Van broeckoven, Christine, & Dhondt, André A. (1992). Extra-pair paternity results from female preference for high-quality males in the blue tit. *Nature, 357*, 494–496.

Lack, D. (1968). *Ecological adaptations for breeding in birds*. Methuen Ltd.

Latour, B. (2005). *Reassembling the social: An introduction to actor-network-theory*. Oxford University Press.

Lawton, M. F., Garstka, W. R., & Hanks, J. C. (1997). The mask of theory and the face of nature. In P. A. Gowaty (Ed.), *Feminism and evolutionary biology, boundaries, intersections and frontiers* (pp. 63–85). Chapman & Hall.

Leonard, Janet. (2019). Hermaphrodite Mating Systems, Encyclopedia of Animal Behavior (Second Edition) 584–590. https://doi.org/10.1016/B978-0-12-809633-8.90133-X.

Lifjeld, J. T., Dunn, P. O., & Westneat, D. F. (1994). Sexual selection by sperm competition in birds: Male-male competition or female choice? *Journal of Avian Biology, 25*, 244–250.

Lindsay, W. R., Andersson, S., Bererhi, B., Höglund, J., Johnsen, A., Kvarnemo, C., Leder, E. H., Lifjeld, J. T., Ninnes, C. E., Olsson, M., Parker, G. A., Pizzari, T., Qvarnström, A., Safran, R. J., Svensson, O., & Edwards, S. V. (2019). Endless forms of sexual selection. *Peer J, 7*, e7988. https://doi.org/10.7717/peerj.7988

Martin, E. (1991). The egg and the sperm: How science had constructed a romance based on male-female roles. *Signs, 16*(3), 485–501.
McKinney, F., Cheng, K., & Bruggers, D. J. (1984). Sperm competition in apparently monogamous birds. In R. L. Smith (Ed.), *Sperm competition and the evolution of animal mating systems*. Academic Press Inc.
Montgomerie, B. (2018). *The invisible women*. American Ornithological Society. https://americanornithology.org/the-invisible-women/
Montgomerie, Bob. (2019). Contemplating the Tundra. https://americanornithology.org/contemplating-the-tundra/.
Nickles, Thomas, "Scientific Revolutions", *The Stanford Encyclopedia of Philosophy*(Winter 2017 Edition), Edward N. Zalta(ed.), URL = https://plato.stanford.edu/archives/win2017/entries/scientific-revolutions/.
Odom, K. J., Hall, M. L., Riebel, K., Omland, K. E., & Langmore, N. E. (2014). Female song is widespread and ancestral in song birds. *Nature Communications, 5*, 3379. https://doi.org/10.1038/ncomms4379
Oring, L. W., Fleischer, R. C., Reed, J. M., & Marsden, K. E. (1992). Cuckoldry through stored sperm in the sequentially polyandrous spotted sandpiper. *Nature, 359*(6396), 631–633. https://doi.org/10.1038/359631a0
Parker, G. A., & Birkhead, T. R. (2013). Polyandry: The history of a revolution. *Philosophical Transactions of the Royal Society B-Biological Sciences, 368*, 1471–2970.
Parker, Patricia G., & Burley, Nancy T. (eds.). (1998). *Avian reproductive tactics: Female and male perspectives*. Ornithological monographs no. 49. Washington DC: The American Ornithologists' Union.
Petrie, M. (1983). Female moorhen compete for small fat males. *Science, 220*, 413–415.
Petrie, M., & Kempenaers, B. (1998). Extra-pair paternity in birds: Explaining variation between species and populations. *Trends in Ecology & Evolution, 13*(2), 52–58.
Ralls, K. (1976). Mammals in which females are larger than males. *The Quarterly Review of Biology, 51*(2), 245–276.
Richards, E. (2017). *Darwin and the making of sexual selection*. The University of Chicago Press.
Richardson, S. S. (2013). *Sex itself: The search for male and female in the human genome*. University of Chicago press.
Røskaft, E. (1983). Male promiscuity and female adultery by the rook Corvus frugilegus. *Ornis Scandinavica, 14*, 175–179.

Ryan, M. J. (2010). An improbable path. In L. C. Drickamer & D. A. Dewsbury (Eds.), *Leaders in animal behavior, second generation.* Cambridge University Press.

Schiebinger, L. (1999). *Has feminism changed science?* Harvard University Press.

Sheldon, B. C. (1994). Sperm competition in the chaffinch: The role of the female. *Animal Behaviour, 47*, 163–173.

Smith, S. M. (1980). Henpecked males: The general pattern in monogamy? *Journal of Field Ornithology, 51*, 55–64.

Smith, S. M. (1988). Extra-pair copulations in black-capped chickadees—The role of the female. *Behaviour, 107*(1–2), 15–23.

Smith, S. M. (2003). Reflections on a lifetime of ornithological research—The 2000 Margaret Morse Nice lecture. *Wilson Bulletin, 115*(1), 73–90.

Snyder, B. F., & Gowaty, P. A. (2007). A reappraisal of Bateman's classic study of intrasexual selection. *Evolution, 61*, 2457–2468.

Stutchbury, Bridget J. M., & Neudorf, Diane. (1998). Female control, breeding synchrony, and the evolution of extra-pair mating systems. In: Parker, Patricia G. & Burley, Nancy T. (eds.) *Avian reproductive tactics: female and male perspectives.* Ornithological monographs no. 49. The American Ornithologists' Union; Washington DC.

Taylor, M. L., Price, T. A. R., & Wedell, N. (2014). Polyandry in nature: A global analysis. *Trends in Ecology & Evolution, 29*(7), 376–383.

Tuana, N. (2006). The speculum of ignorance: The Women's health movement and epistemologies of ignorance. *Hypatia, 21*(3), 1–19.

Veen, T., Borge, T., Griffith, S. C., Sætre, G.-P., Bures, S., Gustafsson, L., & Sheldon, B. C. (2001). Hybridization and adaptive mate choice in flycatchers. *Nature, 411*, 45–50. https://doi.org/10.1038/35075000

Wagner, R. H. (1991). Evidence that female razorbills control extra-pair copulations. *Behaviour, 118*, 157–169.

West-Eberhard, M.-J. (1983). Sexual selection, social competition, and speciation. *Quarterly Review of Biology, 58*, 155–183.

Westneat, D. F., & Stewart, I. R. K. (2003). Extra-pair paternity in birds: Causes, correlates and conflict. *Annual Review of Ecology, Evolution and Systematics, 34*, 365–396. https://doi.org/10.1146/annurev.ecolsys.34.011802.132439

Zuk, M. (2002). *Sexual selections: What we can and can't learn about sex from animals.* University of California Press.

# 5

# Shifting Perceptions about Female Insects, Snakes, Frogs, Lizards, Fishes and Spiders

Marauding locust swarms that destroy crops are known since ancient times in Egypt and India, as well as mentioned in the Iliad, the Bible, the Mahabharata and the Quran. In western North American rangelands, the cyclical occurrence of swarms consists not of locusts (grasshoppers of the family *Acrididae*), but Mormon crickets, *Anabrus simplex* (taxonomically a katydid, not a cricket). They are named after the Mormon settlers, who were allegedly saved by the miracle of seagulls eating the swarms that threatened to destroy one of their first harvests. These Mormon crickets were studied by entomologist[1] Darryl Gwynne, who showed that females, just as males, can also compete for mates—thus opening up for perceptions of females as active and competitive.

This chapter analyses how knowledge/ignorance about females was produced in the different subfields in sexual selection concerning, in turn, insects, snakes, frogs, lizards, fishes and spiders. My analysis is based on interviews and correspondence with researchers from these fields, as well as the scientific record. The researchers interviewed for this chapter are entomologist Darryl Gwynne, herpetologists Richard Shine, Jesús

[1] Entomologists are scientists studying insects.



M. Ah-King, *The Female Turn*, https://doi.org/10.1007/978-981-19-7161-7_5

Rivas and Mike Ryan, fish researcher Charlotta Kvarnemo, and spider researcher Marie Herberstein. The analysis shows what formed the researchers' particular situated knowledges[2] of females as active or multiply mating, or not. Thereby, the analysis demonstrates different ways in which ignorance has been produced.

The particular histories of these subfields show that knowledge about active females and female multiple mating is not always driven by an influx of women scientists in a field combined with gender-reflective interventions to change perceptions about females. There are many divergent routes leading to knowledge about female sexual agency.

## Insect Research—Early Knowledge about Female Multiple Mating

Entomology, the scientific study of insects, has its own particular history when it comes to perceptions about females. It was pest-control studies, natural history observations and population genetics that provided the early evidence that females were mating with multiple males.

Agricultural entomologists had extensive knowledge about insect mating systems because of their work on controlling insect pests. Since the 1950s, they were using the sterile male technique to reduce pest populations, by releasing a large number of sterilized males into the wild to mate with wild females, but the technique only works in species in which females mate with only one male, not if females mate with several males. Therefore, these agricultural entomologists knew that in most insect species females mate with multiple males. In the late 1970s, one group of entomologists in Florida became interested in studying sexual selection. In my interview with Canadian entomologist Darryl Gwynne, he recalls going to the symposium on "the Sociobiology of sex" by the Florida Entomological Society in Gainsville in 1978. He was surprised at how conservative the meeting was—they even prayed at the banquet. Compared to what he was used to, Gwynne says there was a "much more conservative ethic among the Florida entomology society", and he thinks

[2] Haraway, 1988.

"it's almost ironic" that these new ideas about sexual selection emerged among people working in that conservative and religious context.

The group in Florida included firefly researcher James Lloyd, cricket expert Tom Walker and entomologist Randy Thornhill, who was a post-doc in Florida and had written a pioneering paper on females influencing sperm use in 1976 (see Chap. 6). At the 1978 symposium, Lloyd presented the paper "Mating behavior and natural selection",[3] expressing assumptions of active females: "After sperm has been placed in a female, she should manipulate it: store, transfer (from chamber to chamber), use, eat or dissolve it, as she makes additional observations of males."[4] Gwynne recalls that Lloyd's work (as well as Tom Walker's, who studied cricket song) focused on female choice to mate with the right species, but they had told Gwynne that their reading of George Williams' book led to a revolution in their thinking about female choice in terms of individual selection. Working on pre-mating female choice in fireflies *Photuris* species (and firefly taxonomy), Lloyd perceived females as active early on (e.g., he had described how female fireflies mimic the light signal of other species to attract males and catch them as prey[5]).

Independently of the pest-control studies, population geneticists had studied female choice in fruit flies, *Drosophila sp.*, in the 1960s in the laboratory, and found that females often mate with several males.[6] In addition, already in 1974, they used methods to test parentage in fruit flies, through studying chromosomal inversions and found evidence of multiple paternity.[7] These diverse studies showing female multiple mating, along with natural history observations, formed the basis for entomologist and theoretical biologist Geoff Parker's review on sperm competition in 1970,[8] which expanded the study of sexual selection to occur also after mating (post-copulation). Chapter 6 details the history of shifting perceptions from passive toward active females in post-copulatory sexual selection research.

---

[3] Lloyd, 1979.

[4] Ibid: 22.

[5] Lloyd, 2018.

[6] Wyatt Anderson, Personal Communication 2019.

[7] Anderson, 1974.

[8] Parker, 1970.

This history of entomology illustrates production of ignorance about females in sexual selection—although entomologists, natural historians and population geneticists knew about multiple mating in female insects early on, it was not generally acknowledged in the wider field of sexual selection until the 1980s (see Chap. 4).

## Darryl Gwynne—Experimenting with "Gender Fluid" Insects

I met entomologist Darryl Gwynne, a Professor Emeritus in biology, in his office at the University of Toronto Mississauga, Canada. He has a PhD from the Colorado State University (1979) and he describes himself as "a naturalist at heart, I love to go watch bugs in the field". His pioneering studies of Mormon crickets (a species of katydid insect) are commonly presented in textbooks on animal behavior, and opened up for new perceptions about females as active and competitive.

Gwynne notes that he was lucky, he realized that the insects in his parents' backyard would be a perfect study species to test Trivers' new Parental Investment theory.[9] He started a postdoc with Randy Thornhill (the entomologist who coined the term "cryptic female choice", that is females influencing which sperm fertilizes their eggs, see Chap. 6). Thornhill had recently published a paper on courtship feeding in insects and how that could influence sexual selection, where he suggested it as a system for studying sex differences and "role reversals". (Elsewhere, I and behavioral ecologist Ingrid Ahnesjö have criticized the use of the term "sex roles" in animal behavior—because of its cultural connotations, widely disparate usage and risk of masking diverse behavior under the dichotomous terminology of "reversed" versus "conventional sex roles".[10]) In line with this suggestion, Gwynne was interested in investigating species where males provide a large proportion of their body weights in one spermatophore (a capsule containing sperm). In Mormon crickets, the male provides a spermatophore that includes nourishment during mating

[9] Trivers, 1972

[10] Ah-King & Ahnesjö, 2013.

which the female devours as the sperm are transferred. This nutritional package is called a nuptial gift (yet another anthropomorphic metaphor). A male Mormon cricket loses 20 percent of his body weight in producing one spermatophore and its protein content is important for female reproduction. "My very first day in the field, I saw males rejecting females and females fighting, and it was a eureka moment, predicted—and there it is!" Because of the large male investment, females compete for males and males are choosy, they prefer heavier females,[11] Gwynne calls it a "behavioral sex role reversal". He was very excited—the experiments were a "nice fit to the theory" of Robert Trivers and Angus Bateman, therefore, it was well received in the scientific community and did not meet any resistance. Thus, these studies opened up for perceptions of competitive and active females—at least in some unusual species (Fig. 5.1).

## Flexibility in "Sex Roles"

The katydids showed flexibility in which sex compete for mates and which sex choose mates. Because the katydids can modify how much investment in the offspring is provided by the male versus the female, Gwynne describes them as "gender fluid". Although the term "gender" is generally used for socially constructed femininity and masculinity, in biology it is often used interchangeably with sex. Gwynne says he means "gender" in a "zoological, George Williams' sense",[12] that is females showing classic "male characteristics" such as being competitive, seeking mates and being less choosy. Gwynne demonstrated that population density causes flexibility in mate choice and competition in Mormon crickets. Contrary to the common assumption that the mating pattern is species-specific, these experiments showed variability within species and that both males and females can shift between competition and choosiness in mate choice—challenging prevailing perceptions about both females and males. Although male Mormon crickets produce equally big spermatophores independent of population density, at low densities

[11] Gwynne, 1981.

[12] Williams, 1975.

**Fig. 5.1** Entomologist Darryl Gwynne in his office at University of Toronto Mississauga, Canada

males compete for females and females reject males, while at high densities the "courtship roles" are reversed, that is females compete for males and males are choosy.[13] Further experimental studies showed that increasing food availability shifted "role-reversed" courtship behavior to male competition for females.[14] Moreover, when food supply was scarce, the difference among females in reproductive success was higher and thus sexual selection on females stronger.

[13]Gwynne, 1984.

[14]Gwynne & Simmons, 1990.

## Gwynne's View of Sex Differences

When I asked Gwynne whether he experienced a female turn (I was admittedly unclear about what I meant by that at the time) he answered no, "partly because I'm going to be biased to start with, because I started thinking about sex differences and the control of sexual selection because of natural history sex observations of some really intriguing insect species that are basically gender fluid." Here the study species is a clear influence on the researcher's situated knowledges. He was able to do experiments by varying the parameters of parental investment and test predictions, and he found the expected changes in behavior. Gwynne says that in his own research, in terms of the behavior of males and females—"the genders were alike in some, in many respects, it depends on the circumstances in nature". Because that was the beginning of his research career, he "never thought another way really, it's balanced". Thus, his view of the sexes has not changed, it was imprinted by the variability of the katydids from the beginning as well as by Trivers' gender-neutral formulation of Parental Investment theory—"the sex whose typical parental investment is greater than that of the opposite sex will become a limiting resource for that sex".[15] Furthermore, testing the Parental Investment theory about when the sexes are "equal or reversed in their behavior etc" meant Gwynne paid equal attention to reproductive success and sexual selection among females as well as among males,[16] in contrast to the field in general.[17] Furthermore, several of his studies focus on females—one study concerns sexual selection on a female hearing organ which enables them to locate singing males,[18] one on the effect of the nutritious spermatophore for female reproductive success,[19] and another on female mating swarms in a dance fly.[20] In 2012, Gwynne and Patrick Lorch conclude that female mating failure has been found in insects as a result of strong sexual

[15] Trivers, 1972: 56.

[16] For example Gwynne, 1981, 1984.

[17] See Chap. 3 on Malte Andersson's, 1994 summary of sexual selection studies until 1990.

[18] Gwynne & Bailey, 1999.

[19] Gwynne, 1988.

[20] Gwynne & Bussière, 2002.

selection in females.[21] In Gwynne's work, females are active and competing, and he investigates sexual selection in females as well as in males.

## Gwynne's Situated Knowledges

Darryl Gwynne pioneered the research on how the sexes can shift between competing for mates and mate selectively—depending on their relative parental investment, population density and food availability. The unusually flexible sexual behavior of Gwynne's study species, classical sexual selection theory and his experimental investigations formed his situated knowledges of sex and sexual selection. He did not meet resistance to his research since it corroborated Trivers' Parental Investment theory—which predicts that the sex which provides most parental investment will compete for mates. Thus, the katydids provided a model for perceiving the sexes as flexible and sexual selection as dynamic, opening up for perceptions of females as competitive and active—at least in some exceptional species.

# Snake Research: Rick Shine—Benefits of Female Multiple Mating and Ecological Causes of Sex Differences

Rick (Richard) Shine,[22] Professor at University of Sydney, Australia, is the leading authority of Australian snakes and lizards[23] and has published extensively on sexual selection in snakes, frogs, toads, lizards and turtles. When I interviewed him in his Sydney office, he described himself as a natural historian whose research focuses mainly on ecology and evolution (including sexual selection), and lately, on invasive species biology and conservation. His research has contributed to shifting perceptions about

[21] Gwynne & Lorch, 2013.

[22] A search on web of science shows that Shine has published 90 papers on sexual selection.

[23] Encyclopedia of Australian Science and Innovation, Richard Shine.

females in different ways and highlighted alternatives to sexual selection causing sex differences.

Shine states that in contrast to other subfields, such as bird research, snake research was not driven by Angus Bateman's assumptions of competitive indiscriminate males and choosy females (see Chap. 3), because very little was known about snake biology—most of the species Shine has been working on were previously unexplored. In addition, because snakes are not colorful with obvious visual displays and they mainly communicate through chemical signals—sexual selection in snakes is not obvious to humans. Furthermore, as studies were conducted, male snakes were found to be incredibly selective about who they choose to court and very careful in their allocation of courtship effort—contradicting Bateman's prediction of consistently indiscriminate males.

Shine explains that it is difficult to collect behavioral information about snakes, because they are notoriously hard to follow in the field, they can "turn up and then disappear again and they're just not observable". Shine asserts that snake researchers mainly had a natural history perspective which was "simply trying to tease apart systems in which there were both males and females, and they were doing different things". For example, in investigating sex differences in size, they considered fecundity selection (natural selection) to be important in females and male-male competition affecting male size, concluding that the resulting sex difference in size to be the outcome of both (Fig. 5.2).

## Female Adders Benefit from Multiple Mating

In 1992, at a time when female multiple mating was often assumed to be due to male strategies, Thomas Madsen, Rick Shine and coworkers published a study about female benefits of polyandry in the adders, *Vipera berus,* at Smygehuk (Sweden) in *Nature* (one of the two most high-ranked journals in natural sciences).[24] It was the first detailed study of mating patterns in wild snakes. According to Shine "it wasn't challenging other datasets, it might have been challenging some assumptions, but if you

[24] Madsen et al., 1992.

Fig. 5.2 Herpetologist Richard Shine in his office at University of Sydney, Australia

work on snakes you get used to challenging assumptions, because most of what we thought about turned out to be crap." Hence, from Shine's perspective, there wasn't much ground for controversies over polyandry in snakes. The adders paper provided an answer to the puzzle of why females would mate with multiple males, when one mating would be enough for fertilizing the eggs. Since the Smygehuk adder population was isolated, low genetic diversity (inbreeding depression) had led to the high proportion of stillborn offspring and the study showed that females that mated with more males benefited by getting a lower number of stillborn offspring than females mating with fewer males. That is, mating with multiple males increased the viability of the females' offspring. A subsequent experimental introduction of male adders from other populations was shown to increase offspring viability, corroborating that the Smygehuk population was inbred.[25]

Yet, there were two challenges from the scientific community to the adder study, one by Swedish researchers in Gothenburg, arguing that females were monogamous and the stillbirths were due to chemical contaminants, they set up an experiment that—contrary to their expectations—showed that female adders were indeed mating with several males. The other challenge was made by an Italian researcher which didn't find a correlation between multiple mating and offspring viability in an Italian adder population. According to Shine, it was probably due to the study being conducted on a less inbred (more genetically diverse) population. Shine concludes that there were no real challenges, the results were generally received as an answer to the enigma of female multiple mating, and as evidence of benefits of multiple mating was accumulating in other species, the issue became less controversial.

So, how did this finding of benefits to female multiple mating change Shine's view of females? He says that although all snakes lack male parental care and female cooperation is required for successful mating, he previously assumed that female multiple mating was "simply about the cost of refusal and it simply wasn't worth putting up with all this harassment". That is the case for female North American garter snakes (*Thamnophis sirtalis*), they get "mercilessly harassed and sometimes killed—suffocated by courtship from multiple males" at the dens where there may be hundreds of snakes. "The scent of the mating is the cue that switches off courtship by other males and allows the female to escape from the den, that's essentially the only way in which a female can escape from the very large dens, when she's mated." That's not the case for adders, "and by inference that's probably general", reasons Shine. "I guess it changed my worldview", reflects Shine, but he worked a lot on ecological differences between the sexes, and from his perspective, there was never "a notion that in some sense the selective forces of males were more important than

[25] Madsen et al., 1996.

those on females". So, Shine didn't shift from a male-focused perspective to including female agency, it was rather a shift from viewing female multiple mating as a cost-reducing passive strategy to a benefit-incurring active one.

## Feminist Influences

During the interview, Rick Shine mentioned that he had read Germain Greer's the *Female Eunuch* as an undergraduate, and "I was just shattered at my unthinking sexism, and no doubt I still have all kinds of unthinking sexism." In the adders paper, Madsen and coworkers cite Sarah Blaffer Hrdy's chapter "Empathy, polyandry and the myth of the coy female" from Bleier's (1986) *Feminist Approaches to science,* as one of the references showing that polyandry is common. I was wondering whether they kept up with the writings of feminist scientists or primatologists, and Shine says it probably was the primatology literature. Yet, Shine's gender awareness has influenced his scientific publications.

A study on male snake courtship in garter snakes led to a discovery that shifted Shine's understanding of the role of males and females in copulations. It was well-known that it is not possible for a male to force-copulate a female snake, and that males slowly "massage" females during courtship, which Shine describes as "very slow and sensual and it looks very co-operative, and it's always been interpreted as the male stimulating the female and bringing her to some level of receptivity". Yet, at the dens of garter snakes, females that looked healthy were sometimes killed by the mass of males, and they seemed to have been suffocated. Therefore, Shine and colleagues investigated the physiological implications of the massaging behavior and they found that the male "massage" was slowly suffocating the female, causing the reflex to open her cloaca and thus enable intromission. The resulting paper was called "Cryptic forcible insemination: Male snakes exploit female physiology, anatomy, and behavior to obtain coercive matings"[26] (2003). Shine says that it was not easy to write

[26] Shine et al., 2003.

about this finding, "it's very tricky, I remember my collaborator Bob Mason was really nervous about how our female colleagues in particular might respond to that story about forceful insemination". So, to avoid offensive language and check if the writing set off alarm bells, they consulted their female colleagues about the presentation of the results. "It's clearly not rape, rape occurs in human societies. It's mostly about power and everything else. This is a different phenomenon. But it's a real phenomenon. And how do you explain it and tell the story about what's happening without appearing to drop into some pretty dark places?" reasons Shine. To critically reflect upon the use of language is one of the gender-reflective strategies that many feminist scientists and science analysts argue for.[27]

This discussion took place in the wake of the controversy around Randy Thornhill and Craig Palmer's 2000 book *The natural history of rape*, and Shine and his colleagues wanted to avoid associations to rape in humans, and to make sure that their description would not be misinterpreted and cause controversy and critique. Shine says it was "uncomfortable on many levels, I love snakes, and that's a pretty awful dynamic that's going on", but as scientists, "our job is to try to find out what's happening and tell the story with as little role for our own values and so on as possible, be absolutely objective about all". In this uncomfortable case, some consideration was taken to the social reception of their findings. As a consequence, they did not get criticized for it. Hence, gender-reflective approaches[28] may have influenced a wider array of scientists than is often presumed.

## New Findings on Female Reproduction

Another of Rick Shines' papers demonstrates toad females' opportunities for agency in mate choice. They found that female toads (*Bufo marinus*) can inflate their bodies and become turgid, which makes it more difficult for a male to grip the female for mating.[29] This finding suggests that

[27] For example, Schiebinger, 1999.
[28] Richardson, 2013: 17.
[29] Bruning et al., 2010.

females have the ability to exercise more control over the identity of their mates than formerly predicted.

Shine also mentions an interesting ongoing debate about parthenogenesis (meaning "virgin birth" in Greek)—that is reproduction without fertilization by sperm—in lizards and snakes. It is becoming increasingly clear that female lizards and snakes may sometimes opt to reproduce without the involvement of a male. For example, the female Komodo dragon (*Varanus komodoensis*) at London Zoo laid eggs that hatched without having mated with a male. Such findings were initially ignored in herpetology, because the early reports came from people working at zoos and were published in Zoo-type journals with low impact. Reproductive biologists showed some skepticism to those sources, but the cases are accumulating and parthenogenesis among lizards and snakes is increasingly recognized. Thus, this disregard is an example of the production of ignorance through undermining the authority of certain knowers,[30] similar to the western primatologists' former ignorance of the findings of active females by Japanese primatologists (see Chap. 2).

Shine calls attention to another common bias in snake reproductive biology—"Paradoxically, one of the most important aspects of female reproduction in snakes may be that in many populations, most females do not breed in most years".[31] This pattern has resulted in snake researchers focusing on studying "atypical systems with high reproductive frequencies" possibly affecting the overall perception of snake reproduction.[32] Thus, the choice of species to study, and the practical reasons for doing so, may produce bias in the knowledge acquired.

## Ecological Causes of Sex Differences

Sexual selection is not the only cause of evolved sex differences, another cause is ecological factors, if males and females use resources differently, they may evolve in different directions. When Rick Shine reviewed the

[30] Fehr, 2007: 105.
[31] Shine, 2003: 996–7.
[32] Ibid: 996–7.

literature for ecological sex differences in feeding structures, he anticipated to find only a few cases, but he found many examples throughout the animal world. Males and females often evolved to have very different ecological lives—"not regardless of sexual selection, but pretty much peripheral to sexual selection" notes Shine. Because there are so many examples of ecological sex differences of feeding apparatuses, Shine argues that sexual differences in size, which are most often attributed to sexual selection, may well also be due to ecological causes.[33]

Although Charles Darwin already suggested "habits of life" as a cause for sex differences (in parallel with sexual selection), this hypothesis has been overshadowed by the success of the sexual selection approach.[34] Shine suggests that the "ecological causation" model is unpopular compared to the sexual selection hypothesis because the latter is the simplest and most likely to generate falsifiable predictions and hence easier to test. In addition, I suggest there is another reason too, namely the social aspect of science, that because there are trends in which research questions get asked and funded, and sexual selection has been a significant trend in contemporary evolutionary biology—which is associated to the ever-present controversies over sex differences in society.

## Shine's Contribution

Thus, Rick Shine has in many ways contributed to shifting perceptions about females—from passively accepting to mate with courting snake males to active strategies of multiple mating for their own benefits, showing that female toads have capabilities to actively deter courting males, and revealing that the seemingly sensual courtship of male snakes can function to force females to open their cloaca for mating. Furthermore, he has highlighted the common ecological causes of sex differences—in contrast to predominant sexual selection explanations. Shine's situated knowledges are shaped by his natural history observations of snakes and

[33] Shine, 1989:419.

[34] Ibid: 420.

other species in the field, exploring unknown species, and an unusually broad taxonomic overview as well as his gender-reflective approach.

## Jesús Rivas—Disclosing Widespread Female Multiple Mating among Snakes

Venezuelan-born herpetologist Jesús Rivas is a Professor at the New Mexico Highlands University, USA. He is an expert on green anacondas (*Eunectes murinus*) and has contributed to shifting perceptions of female multiple mating in snakes. In contrast to Richard Shine's description of snake biology as without established predictions about mating systems, Rivas provides a different perspective.[35] Belonging to a younger generation of herpetologists, the field was not a blank slate when Rivas entered. When he started studying green anacondas in the 1990s, snakes were generally thought of as polygynous, that is males mating multiple females. It was argued that female multiple mating in snakes was "impeded by phylogenetic reasons", meaning constraints due to evolutionary history. Because many lizards mate polygynously, and lizards are closely related to snakes, researchers assumed that polygyny was the ancestral condition in snakes—a review from 1993 classifies all mating systems found in snakes as different kinds of polygynous systems.[36]

In contrast, Rivas' study of green anacondas indicated that females mate with multiple males. He found breeding balls of one female and up to 13 males, and in one case 17 different males visited a single female during her receptive period. When Rivas was following radio transmitter-marked snakes in the field, he never saw a male courting several females. Furthermore, only half of the males were even able to find a female. It was the first discovery of a polyandrous mating system in snakes and he considered it to be an exceptional mating system among snakes. However, when he and his collaborator Gordon Burghardt made a critical review of the literature, they found that breeding balls with multiple males courting one female were not uncommon. Furthermore, there was frequent

[35] Jesús Rivas, Personal communication in e-mail 2018.

[36] Rivas & Burghardt, 2005: 448.

empirical evidence of multiple mating by females but little evidence of multiple mating in males. Yet, some researchers concluded that their system was polygynous—against their own evidence. Rivas and Burghardt point out that the standard of proof was much higher for multiple mating in females than in males—"unequivocal evidence of copulation with multiple males has not been enough to even suggest that females may be polyandrous, whereas scant field evidence of multiple mating in males has been grounds to conclude that they are polygynous."[37] Hence, the double standard of evaluating evidence for multiple mating in female and male snakes meant that female multiple mating was ignored. Among snakes, the term polyandry was first used to describe the mating system in anacondas, but it is not the only nor the first case documented—just previously unrecognized.[38] Thus, this case exemplifies production of ignorance through the disregard of previous evidence. Rivas and Burghardt suggest that the male bias among snake researchers (8% women first authors of publications during 1973–1982 increasing to 15% 1983–1993)[39] may have been the cause of androcentric interpretations.[40] Furthermore, the constrained perspective of researchers with a certain limited experience of snakes in the northern hemisphere may have biased the interpretation of snake mating systems. Early snake biologists were mostly from temperate zones (in North America) and therefore did not recognize that low encounter rate with potential mates is common among snakes generally, because they had observed masses of snakes at the winter refuges.[41] So, their interpretations of mating systems were influenced by their situated knowledges from a certain geographical location, focusing on the specific breeding system common in temperate snakes which is not representative of snakes in general.[42] Thus, both androcentrism and geographical bias may have contributed to the disregard of female multiple mating in snake research.

---

[37] Ibid: 450.

[38] Ibid: 449.

[39] Wilson, 1998:S35.

[40] Rivas & Burghardt, 2005:450.

[41] Ibid: 447.

[42] Ibid:450–51.

Reviewing the snake literature led Rivas and Burghardt (2005) to reinterpret the role of females in snake mating systems and disclose that female snakes generally mate with multiple males. Among the species breeding in aggregations, eight of the studied species exhibited females mating with multiple males, while in only two species males were shown to mate with multiple females. Furthermore, an accumulating number of studies have shown multiple paternity in snakes. Yet, Rivas and Burghardt note that the term polyandry seems to have been avoided until recently.[43] In snakes, just as in birds, the use of molecular tools to determine the paternity of the broods was important for changing the views on female multiple mating. Although there existed scattered reports of female multiple mating, this mating pattern was only recognized after molecular studies had shown multiple paternity in garter snake litters—yet, garter snakes were still regarded as polygynous[44] (in 2005).

Rivas and Burghardt assert that female multiple mating is common in snakes because female snakes are larger than males (if males are monopolizing several females, they would be expected to compete over females and be larger); most female snakes breed with long intervals (every second year at best), so males ready to breed outnumber females in most cases, which makes it even harder for males to monopolize several females; there needs to be a high availability of females for polygyny to occur (but the common pattern is that it is very hard to find breeding females except for the well-studied breeding aggregations in North America); and males often invest heavily in courting a single female. They also argue that optional parthenogenesis and long-term sperm retention is common in snakes, adaptations suggesting that low encounter rate is common and has a long evolutionary history. "We think that virtually all detailed studies of snake reproductive behavior, viewed objectively, show that snakes' reproductive biology is more consistent with polyandry than with polygyny."[45] Though polygyny in the strict sense (as used for mammals and lizards), namely "a few males monopolize access to many females or in some way manage to gain the benefit of matings with multiple females,

[43] Ibid: 450.

[44] Ibid: 450.

[45] Ibid: 449.

relegating most males in the population to bachelorhood"[46]—might not even be applicable to snakes.[47] Thus, they conclude that polygyny does not describe the predominant pattern among snakes, instead polygynandry (i.e., both males and females mating with multiple mates during the same mating season) is most common, and there are some species with polyandry. In other words, ignorance of female multiple mating has been produced through androcentric interpretations and disregard of empirical evidence in the description of mating systems.

However, new discoveries or interpretations in science do not spread easily; they are often met with resistance, and so was this one. Rivas and Burghardt write "We have presented data implicating polyandry in snakes in recent years at scientific meetings and conferences [...] to skeptical audiences".[48] Rivas recalls the resistance when trying to publish their 2005 review, they tried to send it to all the herpetology journals, but were turned down at each one. Probably the reviewers were the very same authors that Rivas and Burghardt were criticizing for having missed out on female multiple mating although their data suggested it. Thus, Rivas and Burghardt received "a long list of very unfriendly reviewers". Though Rivas recalls that Richard Shine was "one of the few that, to his credit, made constructive comments and supported its publication even if he was not fully convinced at the time". In his review, Shine wrote along the lines that "this argument goes against what we understand of snake mating systems but the authors present a new interesting perspective worth exploring".[49]

Rivas suggests that the cause for this resistance to accept polyandry among snakes as well as to get it published in herpetology journals is "the male bias among snake biologists". He thinks that the theoretical assumption that snake mating systems were constrained to polyandry hindered alternative interpretations. "I believe that unconsciously researchers are more likely to find what they are looking for and what they expect to

[46] Ibid: 447.

[47] Ibid: 450.

[48] Ibid: 449.

[49] Jesús Rivas, Personal communication, e-mail 2018.

find." He assures that the studies in question are good and well reported, "they just missed an aspect of the interpretation".

Since the resistance of researchers within the herpetology field prevented the publication in herpetology journals, it was not until they tried to publish in a journal with a broader theme using a broader list of reviewers, the *Journal of Comparative Psychology*, that they succeeded. Thus, resistance can be confined to a specific sub-field, there can be a consensus among the researchers that makes it hard to even present new ideas. Rivas and Burghardt's paper was published at last, and the perception of female snakes as multiple mating is spreading. Among the 42 studies citing their paper,[50] it is often used as a reference for snakes being polygynandrous or polyandrous (14 papers), four papers use the term promiscuous[51] or polygamous (one study incorrectly cites Rivas & Burghardt 2005 as claiming that "polygyny is the dominant mating system in snakes"[52]) and only two papers present their paper as an alternative interpretation to a consensus, as in the sentence "most snakes are polygynous [reference], but see Rivas & Burghardt 2005". Hence, it seems like perceptions of snake mating systems are shifting.

## Human Bias against Other Species

Apart from revealing ignorance of female multiple mating, Jesús Rivas and Gordon Burghardt have also been engaged in debates about other biases in science. In a chapter entitled *Crotalomorphism: a metaphor for understanding anthropomorphism by omission,* they urge researchers studying animals to "constantly remind ourselves that we are using our human senses and human-based technology, and are processing the information with a human brain".[53] Crotalomorphism, they describe as "the failure to

[50] Number of citing papers according to Web of Science as of January 19 2021.

[51] Rivas & Burghardt (2005: 450) explain that they prefer the term polygynandrous rather than promiscuous, because the latter implies indiscriminate mating, which does not have to be the case when mating with multiple partners.

[52] Aldridge et al., 2020:232.

[53] Rivas & Burghardt, 2002: 9.

consider that other animals have a different world than ours"[54] or "anthropomorphism by omission", which leads to erroneous anthropomorphic interpretations of other species. They write "when we analyze the behavior of animals, we not only can but *must* deliberately put ourselves in the animal's shoes;[55] not doing so is potentially and truly anthropomorphic".[56] The failure of understanding animals is illustrated with many examples, whereof one is a study they themselves carried out investigating the effect of warning coloration on garter snakes. They were close to making the incorrect interpretation that garter snakes reacted to warning (aposematic) colors—but recent research show that snakes cannot distinguish wavelength, that is colors—instead they may have reacted to contrast,[57] the differences between light and darkness. The large focus on warning coloration is due to researchers as humans being very vision-oriented. In their reflections about studying animal cognition, Rivas and Burghardt acknowledge that we can never fully understand how another experiences the world, but "Partial knowledge is possible and useful even if full knowledge is unobtainable both in practice and in principle".[58] Hence, our knowledge of other animals is constrained by our human senses and ways of experiencing the world—that is, with Donna Haraway's terminology, situated.

Thus, Rivas became an advocate for female multiple mating through his discovery of polyandry in green anacondas which lead to his reinterpretation of mating systems in other snakes. Again, the study species was crucial for this researcher's perspectives, motivating his critique of the omission of female multiple mating in snakes. He is also active in calling attention to other biases in science—the geographical bias of focusing on snakes in temperate zones, and the bias of "unthinking anthropomorphism" constraining our understanding of other animals. The early omission of female multiple mating and the scientific community's later resistance to acknowledging female multiple mating illustrate the

[54] Ibid: 10.

[55] It seems difficult to even criticize anthropomorphism in animal behavior studies without using anthropomorphic metaphors!

[56] Rivas & Burghardt, 2002: 11.

[57] Ibid: 12.

[58] Ibid: 11.

cultivation of ignorance through the theoretical focus on males in sexual selection research.

## Frog Research—No Controversy over Female Multiple Mating

Herpetologist Mike Ryan says he doesn't remember any controversy over polyandry outside of the bird research community—"you're not likely to get polyandry in most species of frogs, just because of the way they mate, the male hops onto her back, releases his sperm and she releases eggs, so you're not going to see a lot of polyandry, but there are cases where it happens".[59] He recalls the first publication of female multiple mating in the red-eyed tree frog, *Agalychnis callidryas*, it was considered an interesting finding but did not raise any controversy. According to Ryan then, most female frogs mate with one male (at a time) and there are only a few species of polyandrous frogs, so there has not been a distinctive shift in perceptions of female multiple mating.

On the other hand, entomologist and frog researcher Leigh Simmons points out that DNA technique paternity analyses in frogs as well revealed a greater degree of polyandry than previously acknowledged.[60] There are some species, such as the African foam-nesting tree frog, *Chiromantis rufescens*, in which a female lays eggs with several males in a foam nest. Simmons agrees that the general pattern in frogs and toads is "a male coming together with a female and spawning in a pond", but recently more and more cases show multiple spawning events, in which multiple males are simultaneously holding a mating grip on a single female. Furthermore, even though traditionally males have been thought to fertilize the eggs of the female he is holding in a mating grip, frogs that are not directly involved in the clasping of a female may release sperm into the water in the vicinity of a mating pair and gain fertilizations, just as among spawning fish. Since frogs fertilize their eggs externally (outside of the female body), it has been assumed that paternity is easily quantified,

[59] Mike Ryan interview 2017.

[60] Leigh Simmons interview 2018.

but there is now evidence of mixed paternity clutches. Looking through the literature on frog mating systems, the general expectation given the large sex ratio bias of males at breeding sites was polygyny.[61] Among frogs and toads, polyandry has been reported in at least eight species among five families[62] (out of 7299 species[63]). A recent paper asserts that genetic analyses of mating systems have only been conducted in less than twenty species of frogs and toads, yet demonstrating their diverse mating systems—from extreme monogamy, polygyny to polyandry and polygynandry[64] (both males and females have multiple partners during a breeding season).

Thus, although there was no controversy around polyandry in frog research, paternity analyses have revealed more "extra-pair" paternity and more intense sperm competition than formerly acknowledged, but the perception of female multiple mating has not shifted much.

## Mike Ryan—Understanding Frog Mate Choice "through the brain of the female"

Herpetologist Mike (Michael) Ryan is a Professor of Zoology, and I interviewed him in his office at University of Texas, Austin. His research pioneered experimental evidence of female choice and contributed to the acceptance of female choice as a mechanism of sexual selection (see Chap. 3). He describes his career as "an improbable path", being the oldest of 11 and the son of a truck driver at a time when you were expected to follow in your father's footsteps.[65] Brought up in Bronx and later in rural New Jersey, nature and animals were always his interest. He explains the conception of active females in his research: "what drove me to emphasize the female was—even though I'm not a real neurobiologist—knowing enough about the brain, convinced me very quickly that we had to

[61] Males mating with multiple females over a breeding season.

[62] Byrne & Whiting, 2008.

[63] https://amphibiaweb.org/amphibian/speciesnums.html 2021. New species are discovered every year.

[64] O'Brien et al., 2018.

[65] Ryan, 2010: 465.

understand how the brain was perceiving its world and specifically sexual world." Before his PhD-studies, Ryan did a master thesis (1976–1977) studying territorial behavior in bullfrogs, *Lithobates catesbeianus* ("the males have these wonderful resonating calls, they sound like 'wroomw-room', with bright yellow throats"). At the time, species recognition was the prevailing explanation for courtship signals and variation in signals was generally perceived as noise, but Ryan found that he could tell the difference between individual males only by listening to them and thought "Well, if I can tell the difference, can females tell the difference and do they care?" This question came to guide Ryan's research.

Ryan recounts that what made him open his eyes to the importance of female choice earlier than some others was that, as he started working at Cornell University to conduct his PhD-studies in Neurology and Behavior (1977 to 1982), he had the well-known neurobiologist Bob (Robert) Capranica as a colleague down the hall. Capranica was an electrical engineer from MIT and his research was focused on how animals extract meaningful information from the background noise. His interest was species recognition and he studied "what it was about the frog's brain that guided females to male calls of the same species". Ryan, who was interested in mate choice and recognition *within* species, discussed it with Capranica, who replied that although females will identify male calls of their own species there will be individual variation in which calls they find attractive. Ryan recalls that this clarified that if you want to understand courtship signals, "you had to look at it through the eyes, or more precisely, the brains of the female"—which brings with it "a whole different suite of questions". For example, how do cognitive biases affect female preferences? In addition, Steve Emlen, a renowned expert in bird social systems and one of the founding figures of behavioral ecology, was also at the same department at Cornell, and became an important influence on Ryan's research (Fig. 5.3).

In order to explore the question of variation in female perception of mating calls, Ryan started doing field work on túngara frogs, *Engystomops pustulosus*, in Barro Colorado Island, Panama. Male túngara frogs attract females by sitting in puddles and making "whine chuck chuck" calls—they may use only the whine, but females prefer when the call includes chucks. Through experiments with synthesized calls, Ryan found that

**Fig. 5.3** Herpetologist Mike Ryan in his office in Austin, University of Texas, USA

females prefer larger males and can distinguish their calls by the fundamental frequency (the frequency at which the entire wave vibrates) of the chuck call (the lower the frequency, the larger the male) and accordingly, larger males gain higher reproductive success. The resulting paper was pioneering in showing experimentally that female choose males based on variation in courtship characters in 1980,[66] therefore it was published in the prestigious journal *Science*. Already as a PhD-student, Ryan had developed connections with leading researchers in sexual selection and herpetology, both through his department and through people he met at the Smithsonian Institute of Tropical Biology in Panama, and several of them had read his paper before he submitted it to the journal.[67] The paper was well received by the sexual selection community, as an

[66] Ryan, 1980.

[67] Ryan, 2010:476.

empirical test of Trivers' suggestion that female choice is important, Ryan recalls that only some frog communication researchers were skeptical, because they didn't think that females were attending to this type of variation.

## Sensory Exploitation and Female Agency

As a continuation of this work, Mike Ryan and his coworkers developed the idea of sensory exploitation[68]—that the brain, through the way it works, is biased toward preferring certain signals and therefore a new signal that happens to tap into that preference will be selected for. In the case of frog calls among the different túngara species, Ryan and colleagues showed—through integrating experimental studies, neurobiology and evolutionary history—that there are pre-existing female preferences. The chuck sound is preferred even among females in species where males have not evolved this sound, and the reconstruction of evolution of male traits show that they match these pre-existing biases,[69] in accordance with the sensory exploitation hypothesis.

Ryan and coworkers experienced resistance to the idea of sensory exploitation by researchers who perceived sensory exploitation as ascribing maladaptive behavior to females. This controversy has calmed down now, says Ryan, but there are still people that don't like the sensory exploitation approach. However, Ryan does not agree that sensory exploitation ascribes maladaptive behavior to females, with his perspective of understanding male courtship through the brain of the female, he emphasizes that the brain is under selection to do a lot of other things besides sex. Even if there were no cost to males of these elaborate traits (which there are), you still need to know more than females are choosing males that are good survivors "because that doesn't tell you why fish have bright colors and crickets chirp and fireflies flash and mammals have all these amazing pheromones." Ryan thinks that the purely adaptive approach unintentionally denies the significance of female agency, because it

[68] For example, Ryan, 1990.

[69] Ryan & Rand, 1993: 187ff.

focuses all attention to the males and males having "good genes". According to this way of thinking, the females have onus to distinguish the good males that are good for reproduction, which is a very different way of approach to Ryan's—"that females have these different ways in which they view the world, and males are under very strong selection to catch the females". He thinks advertisement is a good term for it, like manipulation. Males are trying out different things, which sometimes work, "a fish might add a spot in a certain place that makes him more attractive and it's not that the females are under selection to prefer males that are more attractive". According to Ryan's way of thinking, "the females get to decide who's attractive." He thinks that it is a critical though subtle difference to the many studies that use the opposite approach which assumes that females are trying to find the most attractive males. Instead, Ryan's understanding is that if a female mates with someone, he's attractive to that female—"We think that the female's perspective is what's driving everything". Thus, because of his specific methodological approach forming his situated knowledges, Ryan puts emphasis on female agency.

## Ryan's Approach

What formed Mike Ryan and his colleagues' research was understanding enough about the brain to apply it to sexual selection (and perceiving mate choice not just as species recognition), and that their experimental method was able to clearly show mate choice: they synthesized calls and let females move toward one or another call. This is a stronger test, explains Ryan, compared to mate choice experiments in fish which is estimated by time in proximity to one or the other fish. This experimental setup enabled them to investigate not just female preference for this or that stimuli, but the female preference landscape. They played around with making several different kinds of stimuli to see which turned out to be more attractive. Ryan explains: "those experiments were really motivated almost by their entertainment, that 'oh this is fun, let's just see how we can tweak her'". Through their experimental setup and probing the range of female preferences, they developed a "feeling for the organism"—just as feminist science philosopher Evelyn Fox Keller depicts geneticist Barbara McClintock's profound understanding of corn—by being open to what "the material has to say to you".[70]

Ryan's interest in variation in mating signals and his combination of neurobiology, animal behavior and evolutionary approach of figuring out perception of mating calls "through the brain of the female" formed his situated knowledges of females as active. This approach was facilitated by his position and collaboration with colleagues at Cornell University and at the Smithsonian Institute in Panama. Thus, Ryan's research has expanded the evolutionary community's understanding sexual selection—concerning how sensory systems form which signals are perceived as attractive—and his work on female choice contributed to shifting perceptions toward recognizing females as active.

## Lizard Research—A Late Shift toward Acknowledging Female Multiple Mating and Active Females

Recently, a debate about polyandry among lizards took place in the journal *Behavioral Ecology and Sociobiology*. Lizard researchers Ambika Kamath and Jonathan Losos (2017) argued that females in *Anolis* lizards have been assumed to mate monogamously (and males with multiple females) because of long-held and poorly supported assumptions about territoriality. Just as in many other subfields (e.g., bird research), there has been a discrepancy between interpretations of mating systems from behavioral studies and the results from DNA analyses of parentage. Recent genetic studies, published in 2002–2014,[71] show evidence of frequent multiple mating in *Anolis* females and males. Yet, according to Kamath and Losos, descriptions of *Anolis* mating system as generally polygynous still prevail. In response to this critique, other lizard research-

[70] Keller, 1983: 198.

[71] Studies from 2002, 2007 and 2014 (Kamath & Losos, 2017:2).

ers claimed that the ignorance of female multiple mating is not due to faulty assumptions about male territoriality in lizards—a phenomenon that is well supported[72]—rather the prevailing perception of lizards as polygynous follows the general prevailing assumptions about passive and monogamous females in the broader field of evolutionary biology and animal behavior.[73] Behavioral ecologist and lizard researcher Judy Stamps points out that the revelation of female multiple mating was relatively late in the lizard research compared to the general literature on mate choice.[74] She suggests a couple of reasons for this delay—it was only recently that anyone thought of conducting paternity analysis in *Anolis* lizards, and it is difficult to assign maternity to *Anolis* eggs because the females bury them scattered one by one.

Just as in many other sub-fields (e.g., research on primates and birds), early observations of female multiple mating in lizards (such as Gordon 1956)[75] were largely ignored and did not influence the prevailing understanding of the mating system.[76] Thus, ignorance has been produced through not citing early studies of female multiple mating. The lengthy ignorance about female multiple mating in *Anolis* lizards depended on the prevailing assumptions about monogamous females as well as these lizards' specific biology.

Perceptions about female *Anolis* lizards have recently, in the 2010s, shifted from presumably monogamously mating with the territorial male with whom they share space, to mating with multiple males. All genetic studies of *Anolis* lizards suggest that they can now be described as generally polygynandrous,[77] both females and males are mating with multiple partners. Furthermore, female agency in mating has also been endorsed, for example, females are acknowledged to mate also with non-territorial males and to store sperm from previous mates.[78] Although relatively late compared to the wider field of evolutionary biology, perceptions of

[72] Bush & Simberloff, 2018, Stamps, 2018.

[73] Stamps, 2018: 1f.

[74] Ibid: 1f.

[75] Cited in Kamath & Losos, 2018: 8.

[76] Kamath & Losos, 2017: 6ff.

[77] Kamath & Losos, 2018; Stamps, 2018: 1.

[78] Bush & Simberloff, 2018: 3.

females in *Anolis* lizard research has also shifted toward female multiple mating and active females.

## Fish Research—Feminists Highlight the Role of Females

Fish are immensely diverse and more variable in sex than any other vertebrate group,[79] they contain species that change sex, various modes of sex determination including genetic and environmental sex determination, and also species consisting of only females or hermaphrodites[80] (including both male and female sex organs in one individual). Compared to other vertebrates, it is common among males to care for the eggs,[81] although the most common behavior among fish is to spread their eggs and not care, which means that in many species, both males and females show considerable variation in behaviors related to sexual selection,[82] more so than in many other animal groups. Trivers' Parental Investment theory (1972)[83] ignited studies of "sex-role reversed"[84] species, in which males invest more than females in parental investment. For example, in pipefish, males brood the offspring in a pouch or on their bodies. These studies showed that females compete among themselves for males, males are choosy and that the access to males restricts female reproductive success.[85] Furthermore, the pattern is more complex than Parental Investment theory predicted—there is mutual mate choice, males as well as females are choosy, and in some species in which males invest more in parental care, they still compete more for mates than females do.[86] Early studies of sexual selection in fish from 1980–1990, initially also focused on competition between large males and female choice of large and colorful

[79] Vertebrates are organisms with spines, that is mammals, birds, reptiles, amphibians and fish.

[80] Ah-King & Nylin, 2010.

[81] Gross & Sargent, 1985.

[82] Ahnesjö et al., 2008.

[83] Trivers, 1972.

[84] See criticism of the term "sex roles" in Ah-King & Ahnesjö, 2013.

[85] Berglund and Rosenqvist, 1990; Rosenqvist, 1990.

[86] Wilson et al., 2003.

males, but some also studied female choice of male egg guarding and a few studies of male choice for female size.[87] Thus, fish research early on exhibited examples of female choice, male choice, and male as well as female competition.

## Shifting Perceptions of Females in Fish Research

Despite the large observed variability in sexual selection in fish, this research also shows examples of male precedence[88] (that research starts with male-centered investigations or explanations and later includes the female-centered equivalents, a kind of androcentrism[89]). When Canadian researcher Deborah McLennan had studied evolution of male breeding coloration in brook sticklebacks *Culaea inconstans*, and related species for years, she discovered that female sticklebacks also shifted color during the breeding season.[90] Although a multitude of sexual selection studies had been conducted to investigate male color ornaments in sticklebacks during fifty years of research, females were generally described as cryptically colored and shifts in female stickleback coloration had been overlooked. This example of male precedence is in line with the other examples of focusing on male ornaments before investigating female ornaments in birds (see Chaps. 3 and 4), and sexually selected traits in general.

The relationship between male-male competition and female preferences has also been revised. Females have often been assumed to prefer mating with the largest males winning male-male competition, thus salmon researchers long presumed that females preferred the large "fighter" males, and avoided the smaller yet sexually mature males denoted "sneakers". However, fish researcher Jason Watters questioned this assumption. He investigated mating in coho salmon, *Oncorhynchus kisutch*, and found that large males are aggressive toward females and coerce them, and several behaviors indicate that females prefer mating and cooperating with "sneaker" males. Watters suggests that the

[87] Andersson, 1994: 135f.

[88] Ah-King, 2022.

[89] Androcentrism is the practice of placing the male point of view in the center of one's worldview.

[90] McLennan, 1994.

understanding of these two types of males should be revised and the large males should instead be called "coercers" and the small ones "co-operators".[91] Thus, this study challenges the common assumption that large fighting males are also the ones that females prefer.

## Feminist Interventions

Evolutionary behavioral ecologist Charlotta Kvarnemo, Professor at Gothenburg University (Sweden) who studies sexual selection, male care, sperm competition and female ornamentation in fish, recalls that when she started her PhD-studies in 1990, there were plenty of women among her colleagues in animal ecology at Uppsala University, Sweden. During those formative years of hers, they were around ten women, whereof some were self-declared feminists, that all carried out fieldwork at the same marine station, supporting and inspiring each other. Among her collaborators were Ingrid Ahnesjö, Gunilla Rosenqvist, Anders Berglund and Amanda Vincent, investigators studying "sex-role reversed" fish species. An ASAB (Association for the Study of Animal Behavior) workshop, which they organized in Glasgow in 1991, called attention to the neglect of female sexual behavior in sexual selection studies and highlighted female perspectives (an often-used term in the field meaning to consider the world from a female point of view, as opposed to a male-biased perspective) in evolutionary biology and animal behavior research broadly. As a result, Gunilla Rosenqvist and Anders Berglund published the Glasgow workshop review "Is Female Sexual Behaviour a Neglected Topic?" in *Trends in Ecology and Evolution* (1992), and Ingrid Ahnesjö and collaborators wrote "The role of females in influencing mating patterns" in *Behavioral Ecology* (1993). Neither of these papers cited the American feminist primatologists nor the American feminist ornithologists, perhaps this is another site for what feminist philosopher Sandra Harding calls "spontaneous" feminist empiricism[92] (arising from "spontaneous consciousness" of feminist researchers in, for example, biology)

[91] Watters, 2005.

[92] Harding, 2006: 237.

among the European fish researchers that organized the Glasgow workshop. Given the variability in sex, sexual behavior and care patterns among fish, and the observed influence of study species for researchers' situated knowledges, it is not surprising that such insights arose among fish researchers.

Furthermore, since 2008, Ahnesjö has organized several workshops and PhD-courses on gender perspectives in evolutionary biology at Uppsala University, and reviewed and criticized the use of "sex roles" in animal behavior in collaboration with me.[93] Lately, in 2018, Ahnesjö and Paula Vasconcelos organized a workshop on gendered assumptions in evolutionary biology in Uppsala, resulting in published guidelines for alleviating gender bias in evolutionary biology.[94] Thus, feminist researchers working on fish continue to actively promote gender-reflective approaches.

## Currently Understudied Issues

Reflecting on shifts in the field, Charlotta Kvarnemo asserts that the topic of male mate choice has "over-ripely" gained more attention (beyond fish) in the last few years.[95] But she still finds that there is a lack of understanding of male choosiness, a view that is corroborated by another fish researcher, Ingo Schlupp, who recently reviewed the understudied sexual selection topics in *Male Choice, Female Competition, and Female Ornaments in Sexual Selection* (2021).[96]

Fish research has provided many empirical examples that females also compete for mates, that males are also choosy, that sexual selection processes can be both variable over the season and occur mutually. Based on their experiences of specific study species, feminist fish researchers have criticized androcentrism in sexual selection and promoted research focusing on females in animal behavior research generally.

---

[93] Ah-King & Ahnesjö, 2013.

[94] Ahnesjö et al., 2020.

[95] Charlotta Kvarnemo interview 2017.

[96] Schlupp, 2021.

## Spider Research—An Idiosyncratic History of Sexual Selection

Professor Marie Herberstein is an expert on spider behavior and ecology working at the department of Biological Sciences, Macquarie University, Sydney, where I meet her for an interview (Fig. 5.4). With her experience of natural history, sperm competition and cryptic female choice studies of spiders, she provides an inside perspective on sexual selection research in spiders. According to Herberstein, this research has a particular idiosyncratic history, since spider females are much bigger and more sedentary than males, they are the easier sex to study and there has always been a strong focus on females in spider research.[97] There are a lot of women in the field of spider research, "it must be attractive to women, or maybe it's not attractive to men—the constant cannibalism of males", jokes Herberstein. Although research in taxonomy (the classification of organisms) and faunistics (relating to the distribution of animals) is male-dominated, she suggests that it might be only in spider *behavioral* research that women are numerous, or perhaps in behavioral research in general.

When it comes to sexual selection in spiders, Herberstein declares that sexual cannibalism, that is females eating their mate during or after copulation, is an obsession in the field, because it is spectacular and gruesome there seems to be a never-ending need to explain female and male perspectives—who does benefit? Furthermore, she thinks early sexual selection research on spiders in the 1990s was overly focusing on the question "why are males so small?" That question is "a bit boring" from Herberstein's perspective, it seems to come from the assumption that males are generally "big and boofy".[98] But this sex difference is not unique to spiders, it is the common pattern in insects and "it's really just this very small group of boring animals that are different—and that's the mammals". Even among the mammals the pattern is not unitary, for example, rodents and bats have bigger females than males. Therefore, Herberstein finds the focus on trying to explain the small male body size in spiders to be preposterous. It could possibly be a male bias, or a bias due to the researchers

[97] Marie Herberstein interview 2018.

[98] Australian English meaning puffed out.

**Fig. 5.4** Spider researcher Marie Herberstein in her office at Macquarie University, Australia

themselves being mammals, trying to explain an odd pattern from their perspective. To Herberstein, the small male size in spiders is not surprising—they don't fight, they mature early, they can only mate twice anyway (due to their particular mating physiology), all they do is produce sperm and transfer it—there is no need to invest a lot in body size.

In a review of sexual selection research on spiders from 2005, spider researcher Bernhard Huber ascertains that a large proportion of this research has been conducted on for spiders exceptional cases—for example, sexual selection on visual and acoustic signals, extraordinary size differences and cannibalism—and are thus biased toward cases which fit our human primary senses of vision and hearing, or are especially startling from a human perspective.[99] In contrast, Huber points out that the majority of spiders are myopic and mute and therefore rely on vibratory, tactile and chemical communication—of which research is just beginning to develop and will need extensive technology. Furthermore, a group of spiders called araneomorph spiders dominate research, while other groups of spiders have been largely ignored.[100] Similar to what snake researchers Jesús Rivas and Gordon Burghardt were saying above, several spider researchers highlight that our perception of these animals is highly biased by our human primary senses, and that spider primary communication channels are vibrations and chemical communication.[101] Being human is an impediment in understanding communication in animals that mainly use other senses.

## A Natural Historian's Perspective on Spider Sexual Selection

Marie Herberstein describes herself as a natural historian at heart, what excites her is "how weird the whole physiology is in spiders—they don't have flagellated sperm, sperm is encapsulated, they can't swim to the female storage, it's all male hydrostatic pressure that needs to pump these ridiculous sperm into tiny tiny tiny little tubes, into the female, and the tubes are only big enough to let one sperm pass at a time. It's just absurd!" The sperm in its capsule has to be pushed through the female's sclerotized[102] duct—as if through a straw—the ejaculate must be precisely viscous enough to drag the sperm with it, and at the same time not to clog up while the friction is huge, explains Herberstein. Furthermore, because of female sexual cannibalism, males have to transfer thousands of sperm

[99] Huber, 2005.

[100] Ibid: 376.

[101] Ibid: 375.

[102] Hardened cuticle, an insect's outer layer.

in seconds. Some spiders die of a heart attack, because they have to build up internal hydrostatic pressure in order to squeeze the sperm out. It is the absurdity that is interesting, says Herberstein, the physiological and morphological limitations of this mating system, "rather than who's winning, who's in control". Spiders are exciting because they have odd behavior, generating lots of questions about how they do it and why they do it. Furthermore, Australia is a great place for a natural historian to study interesting organisms and Herberstein also conducts research on insects, such as mantids and damselflies.

## Discovering Cryptic Female Choice in Spiders

Herberstein describes one of her experiments as "a classic sperm competition experiment" on St Andrew's Cross Spider, *Argiope keyserlingi,* in which females mated with two males, then the fertilization success by each male was determined and related to differences in male size and copulation duration. The only result that turned out significant was that the relative copulation duration affected fertilization success—if the second male copulated relatively longer than the first male, he fertilized more eggs, but only when both males were cannibalized.[103] It's the female that is in control of the copulation duration, because when she eats the male—the copulation ends. Thus, the experiment showed that when the male was smaller, the female allowed longer copulation, therefore (in contrast to classical sexual selection predictions) the smaller males end up fertilizing more eggs. Hence, this study started out as a sperm competition study, but the results were better interpreted as cryptic (or rather overt) female choice.

Herberstein says sperm competition is the default position in post-copulatory sexual selection studies, if the results do not fulfill expectations of sperm competition, cryptic female choice is the second possibility. She ponders whether it is because it is called *cryptic* female choice—what you can't see. "What if it was just called female choice, would we look at it differently? Or are we going—ah, it's cryptic, I can't possibly see it, so

[103] Elgar et al., 2000.

we're just not going to look in the first place. And therefore, we do it by elimination? If you think about it, sperm competition is not obvious." It is difficult to determine which of the two mechanisms affects the outcome of what happens in the "black box" of female genitals. Herberstein adds decisively, "Well, spiders are the solution, obviously. Because you can compare the ejaculate versus how much is being taken up." Male spiders have two organs on their legs in which they store all their sperm, so it is possible to measure how much sperm has been ejaculated, how much is left in the organ, and also compare with the content of the other unused organ—therefore, she explains, it would be possible to discern between male sperm competition strategies and female storage or choice strategies.

## Taxonomic Resistance

To my question whether she has experienced any resistance to her work, Marie Herberstein answers "Only the resistance that all I find about spiders is going to be only relevant to spiders and not anything else. I only find the *taxonomic* resistance." That females have control in spiders has not been controversial. Spiders—and pipefish in which males carry offspring—they are never viewed as good systems to test assumptions, they are treated like "just an odd thing" ascertains Herberstein. In contrast, she thinks they are great models for investigating choice and competition under conditions when males have a higher cost of reproduction. Taxonomic resistance, what some other researchers refer to as taxonomic bias, or in line with the terminology in the epistemology of ignorance, I call taxonomic ignorance, occurs frequently in sexual selection (see next section).

Herberstein's way of understanding sex and sexual selection is formed by her natural history interest, her gender-reflective perspectives on the research focus on small male size, and her studies of the specific biology of spiders—females are large and in control over mating and sometimes even of their mates' immediate survival, that sperm are encapsulated and need to be pressed into the female sperm stores by building up the hydrostatic pressure in the male. Spider research has largely been female-focused, because females are larger and easier to study. Yet, even in Herberstein and colleagues' spider study on post-copulatory sexual selection, sperm competition was the default assumption (as generally in the evolutionary biology field), when they discovered cryptic female choice. Hence, perceptions about female spiders have been active from the start, female multiple mating was studied since 1980,[104] but androcentric sperm competition theory and methodology have influenced research in this field as well (Fig. 5.5).

## Taxonomic Bias

The taxonomic bias in sexual selection studies began with Charles Darwin's *Descent of Man* which relied heavily on his available information on birds, mammals and insects, and when Malte Andersson reviewed the field in *Sexual Selection* (1994), studies were conducted predominantly on birds, insects, amphibians, fishes and mammals[105] (in that order of magnitude).

Furthermore, there is a generally acknowledged taxonomic bias in knowledge among sexual selection researchers. Herpetologist Mike Ryan recalls that when he was a graduate student in the late 1970s, behavioral ecology students were required to know the bird literature, but students working with birds had no academic pressure to know about other animal groups.[106] Similarly, entomologist Leigh Simmons is critical of the asymmetrical taxonomic bias in citations—insect researchers cite broadly the studies of birds and mammals for example, but bird researchers seldom refer to insect research. This pattern is confirmed by ornithologist Simon Griffith, who confesses that "I would apologetically admit that I don't follow too much stuff outside my own taxonomic group."[107]

My interpretation is that this taxonomic bias is associated with the status of birds as the "paradigmatic taxon" of sexual selection (see Chap.

[104] Ibid: 373.

[105] Huber, 2005: 365 represents the relative focus on different animals in illustrative diagrams.

[106] Mike Ryan interview 2017.

[107] Simon Griffith interview 2018.

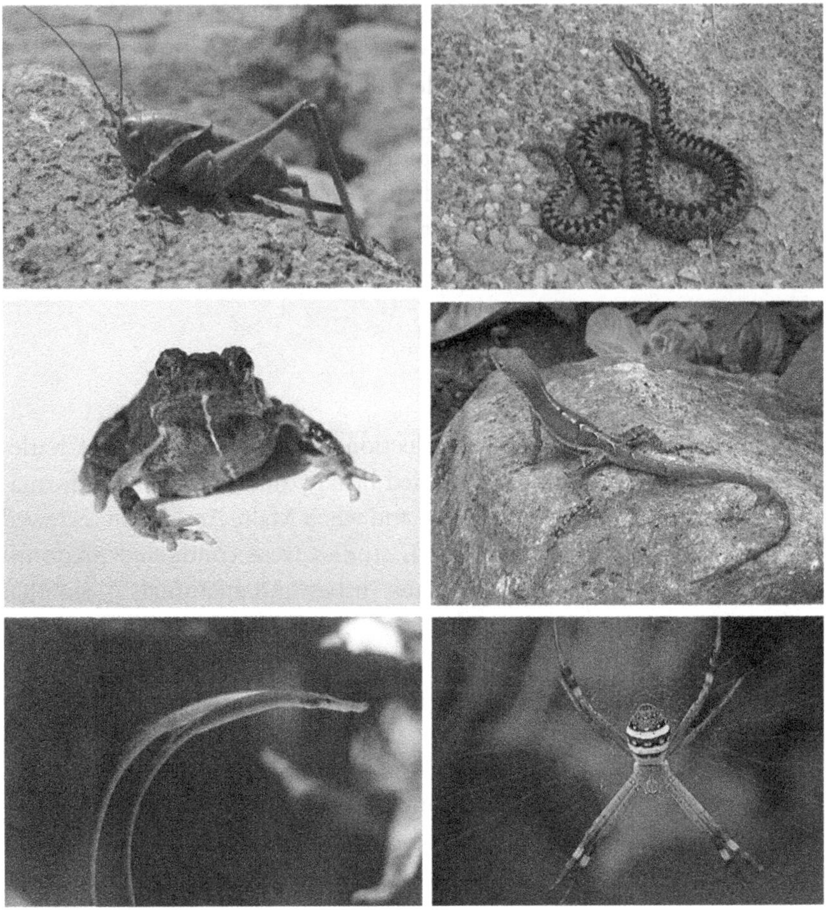

**Fig. 5.5** Photographs illustrating study animals in this chapter. A) Female Mormon cricket *Anabrus simplex,* photo by Jerry Friedman. B) Female adder, Vipera berus, photo by Loch Shin. C) Female tungara frog *Engystomops pustulosus*, photo by Brian Gratwicke. D) Female Watts's tree lizard, *Anolis wattsi*, photo by Gail Hampshire. E) A pair of broad-nosed pipefish, *Syngnathus typhle*, photo by Miguel Mendez. F) Female St Andrew's Cross Spider, *Argiope keyserlingi,* photo by James Niland. Photos cropped and published under Creative Commons Attribution 2.0 Generic except B) which is public domain

4), therefore they are at the top of the reference hierarchy in the field. Although Darryl Gwynne's classic example of flexible mating behavior in katydids is well-known in sexual selection, marine ecologist Ingrid Ahnesjö, who studies mating flexibility in pipefish, expresses her concern that research on sexual selection in "sex role reversed"[108] species is generally discussed as "the exception that confirms the rule".[109] This view corroborates with spider researcher Marie Herberstein's perception that spiders, just as pipefish, are seen as odd examples rather than good models to test general assumptions.

Furthermore, there are some studies of taxonomic citation bias. One study examined publications on parental care and found that studies on birds and mammals cite a relatively narrow range of animal groups compared to studies on reptiles, amphibians, fishes and invertebrates.[110] In addition, fish expert and editor of the journal *Ethology* (the study of animal behavior) Michael Taborsky investigated studies on behavior and showed that bird and mammal studies cite publications in their own animal group to a larger degree than do studies on other animals.[111] He found that the research questions, or where in the world the authors come from, did not affect this bias and concludes that it is likely due to "taxonomic parochialism", that is "the bias to refer to only a narrow range of taxonomic groups when discussing the context of a study". Taborsky pleads to researchers to avoid this taxonomic ignorance and argues for the benefits of taxonomic comparison, namely "the wealth of knowledge, ideas and insight we can gain when considering how different organisms solve similar problems. So often in biology comparison affords amazing surprises, which a blinkered view is going to miss."[112]

There is thus a general acknowledgment of taxonomic bias in knowledge and citations in sexual selection, both in the interviews and in the literature. Birds have been described as the "paradigmatic" animal group in sexual selection, and bird studies are well read and cited among the

[108] See criticism of the term "sex roles" in Ah-King & Ahnesjö, 2013.

[109] Ingrid Ahnesjö Personal Communication 2016.

[110] Stahlschmidt, 2011:3. The study concerned publications from 2001–2010.

[111] Taborsky, 2009.

[112] Ibid: 108.

researchers working on other taxa, but bird researchers admit to not following the research in other taxa that well. This asymmetry in knowledge circulation is also part of the cultivation of ignorance in sexual selection. Indeed, if the subfields in sexual selection had had more knowledge exchange, or more equal status in the field, perhaps there had not been such delays in knowledge production about females?

## Chapter Conclusions

This chapter shows that knowledge about female multiple mating and female agency was reached through many different routes in studies on insects, snakes, frogs, lizards, fishes and spiders—through agricultural pest-control, natural history studies, population genetics, interdisciplinary methodological development, studies of specific animals and feminist interventions—demonstrating that such knowledge is sometimes but not always driven by feminists' gender-reflective interventions to change perceptions and advance knowledge about females. These histories nuance the earlier research in feminist science studies that often focus on fields (primatology,[113] developmental biology,[114] medicine[115] and sex determination research[116]) which have had an influx of women scientists associated with feminist critique and introductions of new paradigms.

In these diverse subfields, the researchers' situated knowledges were formed by their theoretical perspectives, their specific study species, their geographical location, their specific methodologies and occasionally, by feminist insights. Darryl Gwynne maintains that his understanding has all along been that females and males have flexible strategies in mate choice and competition, because he started out his research working on katydids with flexible "behavioral sex roles". Rick Shine describes his shift from viewing female multiple mating in snakes as a cost-reducing passive strategy to a benefit-incurring active one. Snake researcher Jesús Rivas

[113] Haraway, 1989, Schiebinger, 1999, Nelson, 2017, see Chap. 2.

[114] Schiebinger, 1999, Nelson, 2017.

[115] Schiebinger, 1999, Nelson, 2017.

[116] Richardson, 2013.

recounts how his discovery of female multiple mating in green anacondas led to his reevaluation of the prevailing perception of snake mating systems as predominated by male multiple mating. Mike Ryan's interdisciplinary neurobiology-experimental-evolutionary history approach and his questions about the reception of variability in mating calls in frogs were one way in which active females have been put at center stage. In fish research, feminist biologists conducting research on species in which males invest heavily in parental care, criticized androcentrism in sexual selection and actively promoted research focusing on understudied females in animal behavior generally. In contrast spider research has been largely focused on the females, which are larger and easier to study than the males.

Thus, shifts in perceptions of females happened at different times, knowledge about female multiple mating emerged early (1950s and 1960s) in insect research, 1980 and onwards in spider research, its recognition in snake mating systems shifted after 2005 and recently in lizard studies. Since frogs and fishes have mostly external fertilization (there are fishes with internal fertilization), perceptions of female multiple mating have not shifted much, although higher degrees of multiple paternity than formerly known has been revealed by DNA fingerprinting studies.

These subfields' histories show several ways in which ignorance has been produced, through androcentrism, male precedence, undermining the authority of certain knowers, overlooking alternative interpretations and taxonomic ignorance. The degree of androcentrism differs in the early development of these subfields, it seems to have been low in spider research, rather low in snake and frog studies, variable in insect research (see Chap. 6), high in lizard research and there are some examples of male bias in fish research. Some studies explicitly conducted to test sexual selection theory (such as sperm competition) were androcentric, at the same time, entomologist Darryl Gwynne's sexual selection studies of flexible mate choice behavior and competition were early to investigate female competition and sexual selection in females.

Just as in bird research, there was male precedence in fish research, focusing on males' more spectacular ornaments first, which produced ignorance and delayed knowledge production about females' more subtle ornaments.

The reproductive biology research community undermined the authority of knowers publishing in zoo-type journals, which delayed the recognition of optional parthenogenesis (female agency to optionally reproduce without involving a male) in some lizards and snakes.

The snake research community, with its prevailing view that snake mating systems were predominated by male multiple mating, overlooked that female multiple mating is widespread. Due to the resistance among snake researchers, this alternative interpretation was hindered from publication in herpetology journals, but was later accepted in a broader journal.

Furthermore, there is a generally acknowledged knowledge and citation hierarchy in sexual selection in which bird research is on top, followed by insect research, and other animal groups following thereafter. In addition, certain animals, such as spiders, are often interpreted as curious exceptions rather than recognized as legitimate examples that can be used for understanding general patterns in sexual selection. This asymmetric circulation of knowledge leads to taxonomic ignorance in sexual selection—a more symmetric circulation of knowledge might have speeded up the female turn.

## References

Ah-King, M. (2022). The history of sexual selection research provides insights as to why females are still understudied. *Nature Communications, 13*(1), 6976. http://doi.org.10.1038/s41467-022-34770-z

Ah-King, M., & Ahnesjö, I. (2013). The "sex-role" concept: A review and evaluation. *Evolutionary Biology, 40*, 461–470.

Ah-King, M., & Nylin, S. (2010). Sex in an evolutionary perspective: Just another reaction norm. *Evolutionary Biology, 37*, 234–246.

Ahnesjö, I., Vincent, A., Alatalo, R., Halliday, T., & Sutherland, W. (1993). The role of females in influencing mating patterns. *Behavioral Ecology, 4*(2), 187–189.

Ahnesjö, I., Forsgren, E., & Kvarnemo, C. (2008). Variation in sexual selection in fishes. In C. Magnhagen, V. A. Braithwaite, E. Forsgren, & B. G. Kapoor (Eds.), *Fish behaviour*. Science Publishers.

Ahnesjö, I., Brealey, J. C., Günter, K. P., Martinossi-Allibert, I., Morinay, J., Siljestam, M., Stångberg, J., & Vasconcelos, P. (2020). Considering gender-biased assumptions in evolutionary biology. *Evolutionary Biology, 47*, 1–5. https://doi.org/10.1007/s11692-020-09492-z

Aldridge, R. D., Siegel, D. S., Goldberg, S. R., & Pyron, R. A. (2020). Seasonal timing of spermatogenesis and mating in squamates: A reinterpretation. *Copeia, 108*(2), 231–264.

Anderson, W. W. (1974). Frequent multiple insemination in a natural population of Drosophila pseudoobscura. *The American Naturalist, 108*(963), 709–771.

Andersson, M. (1994). *Sexual selection*. Princeton University Press.

Berglund, A., & Rosenqvist, G. (1990). Male limitation of female reproductive success in a pipefish—Effects of body-size differences. *Behavioral Ecology and Sociobiology, 27*, 129–133.

Bruning, B., Phillips, B. L., & Shine, R. (2010). Turgid female toads give males the slip: A new mechanism of female mate choice in the Anura. *Biology Letters, 6*, 322–324. https://doi.org/10.1098/rsbl.2009.0938

Bush, J. M., & Simberloff, D. (2018). A case for anole territoriality. *Behavioral Ecology and Sociobiology, 72*, 111. https://doi.org/10.1007/s00265-018-2522-6

Byrne, P. G., & Whiting, M. J. (2008). Simultaneous polyandry increases fertilization success in an African foam-nesting treefrog. *Animal Behaviour, 76*, 1157–1164.

Elgar, M. A., Schneider, J. M., & Herberstein, M. H. (2000). Female control of paternity in the sexually cannibalistic spider Argiope keyserlingi. *Proceedings of the Royal Society of London Series B, 267*, 2439–2443.

Encyclopedia of Australian Science and Innovation. *Richard Shine* https://www.eoas.info/biogs/P005621b.htm.

Fehr, C. (2007). Are smart men smarter than smart women? The epistemology of ignorance, women and the production of knowledge. In A. M. May (Ed.), *The 'woman question' and higher education: Perspectives on gender and knowledge production in America*. Edward Elgar Publishing Inc.

Gross, M. R., & Sargent, R. C. (1985). Evolution of male and female parental care in fishes. *American Zoologist, 25*(3), 807–822.

Gwynne, D. T. (1981). Sexual difference theory: Mormon crickets show role reversal in mate choice. *Science, 213*(4509), 779–780.

Gwynne, D. T. (1984). Sexual selection and sexual differences in Mormon crickets (orthoptera: Tettigoniidae, Anabrus simplex). *Evolution, 38*(5), 1011–1022.

Gwynne, D. T. (1988). Courtship feeding and the fitness of female katydids (orthoptera: Tettigonidae). *Evolution, 42*(3), 545–555.

Gwynne, D. T., & Bailey, W. J. (1999). Female-female competition in katydids: Sexual selection for increased sensitivity to a male signal? *Evolution, 53*(2), 546–551.

Gwynne, D. T., & Bussière, L. F. (2002). Female mating swarms increase predation risk in a 'role-reversed' dance fly (Diptera: Empididae: Rhamphomyia longicauda Loew). *Behaviour, 139*, 1425–1430.

Gwynne, D. T., & Lorch, P. D. (2013). Mating failures as a consequence of sexual selection on females. *Entomologia Experimentalis et Applicata, 146*, 3–10.

Gwynne, D. T., & Simmons, L. W. (1990). Experimental reversal of courtship roles in an insect. *Nature, 346*, 172–174.

Haraway, D. (1989). *Primate Visions: Gender, race and nature in the world of modern science*. Routledge.

Haraway, D. J. (1988). Situated knowledges: The science question in feminism and the privilege of partial perspectives. *Feminist Studies, 14*(3), 575–599.

Harding, S. (2006). Rethinking standpoint epistemology: What is 'strong objectivity'? In E. F. Keller & H. E. Longino (Eds.), *Feminism and science* (pp. 235–248). Oxford University Press.

Huber, B. A. (2005). Sexual selection research on spiders: Progress and biases. *Biological Reviews, 80*, 363–385.

Kamath, A., & Losos, J. (2017). The erratic and contingent progression of research on territoriality: A case study. *Behavioral Ecology and Sociobiology, 71*, 89. https://doi.org/10.1007/s00265-017-2319-z

Kamath, A., & Losos, J. (2018). Reconsidering territoriality is necessary for understanding Anolis mating systems. *Behavioral Ecology and Sociobiology, 72*, 106.

Keller, E. F. (1983). *A feeling for the organism: The life and work of Barbara McClintock*. W. H. Freeman.

Lloyd, J. E. (1979). Mating behavior and natural selection. *The Florida Entomologist, 62*(1), 17–34.

Lloyd, James E. (2018). A naturalist's long walk among shadows: of North American Photuris—patterns, outlines, silhouettes… echoes. Self-published, Gainesville, FL, USA.

Madsen, T., Shine, R., Loman, J., & Håkansson, T. (1992). Why do female adders copulate so frequently? *Nature, 355*(6359), 440–441.

Madsen, T., Stille, B., & Shine, R. (1996). Inbreeding depression in an isolated population of adders, Vipera berus. *Biological Conservation, 75*, 113–118.

Malin, Ah-King (2022). The history of sexual selection research provides insights as to why females are still understudied. *Nature Communications 13*(1) 6976 10.1038/s41467-022-34770-z

McLennan, D. A. (1994). Changes in female colour across the ovulatory cycle in the brook stickleback, Culaea inconstans (Kirtland). *Canadian Journal of Zoology, 72*(1), 144–153. https://doi.org/10.1139/z94-018

Nelson, L. H. (2017). *Biology and feminism: a philosophical introduction.* Cambridge University Press.

O'Brien, D. M., Keogh, J. S., Silla, A. J., & Byrne, P. G. (2018). The unexpected genetic mating system of the red-backed toadlet (Pseudophryne coriacea): A species with prolonged terrestrial breeding and cryptic reproductive behaviour. *Molecular Ecology, 27*(14), 3001–3015.

Parker, G. A. (1970). Sperm competition and its evolutionary consequences in the insects. *Biological Reviews, 45*, 525–567.

Richardson, S. S. (2013). *Sex itself: The search for male and female in the human genome.* University of Chicago press.

Rivas, J. A., & Burghardt, G. M. (2002). Crotalomorphism: A metaphor to understand anthropomorphism by omission. In M. Bekoff, C. Allen, & G. M. Burghardt (Eds.), *The cognitive animal: Empirical and theoretical perspectives on animal cognition* (pp. 9–17). MIT Press.

Rivas, J. A., & Burghardt, G. M. (2005). Snake mating systems, behavior, and evolution: The revisionary implications of recent findings. *Journal of Comparative Psychology, 119*(4), 447–454.

Rivas, J. A., Muños, M. C., Burghardt, G. M., & Thorbjarnarson, J. B. (2007). Sexual size dimorphism and the mating system of the green Anaconda (Eunectes murinus). In R. W. Henderson & R. Powell (Eds.), *Biology of the boas and pythons* (pp. 313–225). Eagle Mountain Publishing.

Rosenqvist, G. (1990). Male mate choice and female-female competition for mates in the pipefish Nerophis ophidion. *Animal Behaviour, 39*, 1110–1115.

Rosenqvist, G., & Berglund, A. (1992). Is female sexual behaviour a neglected topic? *Trends in Ecology and Evolution, 7*(6), 174–176.

Ryan, M. J. (1980). Female mate choice in a neotropical frog. *Science, 209*, 523–525.

Ryan, M. J. (1990). Sensory systems, sexual selection, and sensory exploitation. *Oxford Surveys in Evolutionary Biology, 7*, 157–195.

Ryan, M. J. (2010). An improbable path. In L. C. Drickamer & D. A. Dewsbury (Eds.), *Leaders in animal behavior, second generation* (pp. 465–496). Cambridge University Press.

Ryan, M. J., & Rand, A. S. (1993). Sexual selection and signal evolution: The ghost of biases past. *Philosophical Transactions: Biological Sciences, 340*(1292), 187–195.

Schiebinger, L. (1999). *Has feminism changed science?* Harvard University Press.

Schlupp, I. (2021). *Male choice, female competition, and female ornaments in sexual selection*. Oxford University Press.

Shine, R. (1989). Ecological causes for the evolution of sexual dimorphism: A review of the evidence. *Quarterly Review of Biology, 64*(4), 419–461.

Shine, R. (2003). Reproductive strategies in snakes. *Proceedings of the Royal Society of London Series B, 270*, 995–1004. https://doi.org/10.1098/rspb.2002.2307

Shine, R., Langkilde, T., & Mason, R. T. (2003). Cryptic forcible insemination: Male snakes exploit female physiology, anatomy, and behavior to obtain coercive matings. *American Naturalist, 162*(5), 653–667.

Stahlschmidt, Z. R. (2011). Taxonomic chauvinism revisited: Insight from parental care research. *PLoS One, 6*(8), e24192. https://doi.org/10.1371/journal.-pone.0024192

Stamps, J. (2018). Polygynandrous anoles and the myth of the passive female. *Behavioral Ecology and Sociobiology, 72*, 107. https://doi.org/10.1007/s00265-018-2523-5

Taborsky, M. (2009). Biased citation practice and taxonomic parochialism. *Ethology, 115*(2), 105–111.

Trivers, R. L. (1972). Parental investment and sexual selection. In B. Campbell (Ed.), *Sexual selection and the descent of man* (pp. 136–179). Aldine.

Watters, J. V. (2005). Can the alternative male tactics "fighter" and "sneaker" be considered "coercer" and "cooperator" in coho salmon? *Animal Behaviour, 70*, 1055–1062.

Williams, G. C. (1975). *Sex and evolution*. Princeton University Press.

Wilson, D. S. (1998). Patterns of publication in three north American herpetological journals: Gender biases. *Herpetologica, 54*(Suppl 1), S35–S42.

Wilson, A. B., Ahnesjö, I., Vincent, A., & Meyer, A. (2003). The dynamics of male brooding, mating patterns, and sex roles in pipefishes and seahorses (family Syngnathidae). *Evolution, 57*, 1374–1386.

# 6

# Sexual Selection After Mating: Turning Perceptions of Female Agency

When a cow defecates on a meadow, at least in the Northern hemisphere, the dung is usually covered in yellow dung flies, *Scathophaga stercoraria* (the common name refers to the male coloring, females are greenish) within five minutes of reaching the ground. The males arrive first, females arrive less quickly than males, and tend to fly over the dung patch landing up-wind and walk back to it.[1] If walking into a field near Langford, UK, in the late 1960s, one might have observed British entomologist[2] and theoretical biologist Geoff Parker walking around tagging each single dung patch with gardening labels, measuring temperature, light conditions, humidity, counting males and females, following the daily rhythm of cattle defecation and laying in the grass on an old coat side-wind to a dropping watching fly behavior.[3] He observed that males fight intensively among themselves to mate with the arriving females, and often try to replace males that are mating with or hang on to the back of a female while she lays eggs. Females do not seem to have much influence over

[1] Parker, 1970a: 140.

[2] An entomologist is a scientist that studies insects.

[3] Parker, 2001: 6.



M. Ah-King, *The Female Turn*, https://doi.org/10.1007/978-981-19-7161-7_6

whom they mate with—most mate with the first male encountered.[4] Though on rare occasions, Parker noted, females face a searching male and "rock from side to side with the wings diagonally outwards from the vertical axis of the body, or flick each wing alternately"[5] as if to catch the male's attention, and this behavior invariably leads to mating. Five hours after the dung dropping, the fuzz and the buzz is over and few dung flies remain.[6] These detailed studies of dung fly behavior led Parker to suggest that, just as individual males can compete, sperm from different males can also compete for fertilizing a batch of eggs.[7] This sperm competition theory expanded sexual selection theory to also include what happens after mating. Originally, Charles Darwin had formulated sexual selection theory around competition and mate choice before mating, but did not conceive of (as far as we know) sexual selection after mating.

Based on interviews and correspondence with Geoff Parker, entomologist Randy Thornhill, tropical ecologist William Eberhard, entomologist Leigh Simmons and ornithologist Simon Griffith, as well as the scientific record, this chapter examines the development and reception of the equivalent male and female hypotheses about sexual selection after copulation: sperm competition and cryptic female choice (females influencing which sperm fertilizes their eggs during or after mating). Why was sperm competition the early focus of post-copulatory sexual selection? Why was the equivalent idea for females—cryptic female choice—developed later than sperm competition and met with skepticism by some investigators? Expanding the realm of sexual selection to occurring also after copulation meant that sperm competition and cryptic female choice were equally relevant hypotheses for influencing resulting fertilizations, but as it has repeatedly been the case in the history of sexual selection research, males were the initial focus. Thus, here is another example production of ignorance about females through male precedence (that research starts with male-centered investigations or explanations and later includes the female-centered equivalents).[8]

[4] Parker, 1970a: 142.

[5] Ibid.: 148.

[6] Parker, 1970c: 198.

[7] Parker, 1970b.

[8] Ah-King, 2022.

Both ideas were developed first in insect research and later applied in other animals. This chapter relates this development first in insect research and then the subsequent expansion to bird research. Throughout this chapter, I investigate what formed the researchers' particular situated knowledges[9] of females as active or not. Thereby the analysis demonstrates different ways in which ignorance has been produced in sexual selection.

## Geoff Parker: Founding Sperm Competition

Entomologist and theoretical biologist Geoff Parker, with a PhD from the University of Bristol (1969), is the influential founder of the sperm competition subfield in sexual selection. He is now a Professor Emeritus and I met him at his office at Liverpool University (Fig. 6.1). He grew up as a natural history kid, with a dad who was a PhD and a great naturalist.[10] In his youth, Geoff Parker says, he was obsessed with animals and best friends with a farmer's son, at whose farm he spent a lot of time, and that's where he first encountered the yellow dung flies. Later, during his undergraduate zoology studies at the University of Bristol, when he had to do his final last year project, he chose insect behavior. Parker decided on yellow dung flies when his supervisor mentioned them, because he was acquainted with them at the farm and he was fascinated by the fights between males. Through his studies, Parker developed yellow dung flies into a model system[11] (a species that is extensively studied and thereby become experimentally powerful to answer particular questions) and it turned out to be a very productive starting point. As Parker explains, "A good model system should make one ask wider questions, I owe to the dung-flies the fact that I became interested in a range of general problems, such as sperm competition, the evolution of two sexes, animal contests, animal distributions and sexual conflict".[12] Once again, as in the

[9] Haraway, 1988.

[10] Geoff Parker interview 2017.

[11] Parker, 2001.

[12] Parker, 2010a: 268.

Fig. 6.1 Geoff Parker in his office at Liverpool University, UK

previous chapters, the study species is an important influence in forming the researcher's situated knowledges.

## Pioneering Sexual Selection Studies

Geoff Parker's motivation for studying dung flies was to explore sexual selection. When he started his research in 1965, the prevailing theoretical framework was group selection (the notion that selection acts on whole groups of organisms, for example as selection "for the good of the species"), and selection at the individual level (the notion that selection acts on individuals and lead to individual adaptation) was a new and controversial perspective (see Chap. 3). As a consequence, sexual selection was not generally accepted as important in evolution.[13] For example, a PhD student in Bristol published a paper on dragonflies in the 1950s, in which he investigated why male dragonflies had territories and he did not even consider sexual selection. His suggested explanations were about the benefits of spacing for the population (such as reduced parasite load), but none of the suggested explanations had anything to do with gaining access to females.[14] Given the prevailing "the good of the species" consensus, Parker says he was glad that his supervisor, entomologist Howard Hinton, did not read his thesis properly, or he probably would not have written the subsequent supportive recommendation letters for Parker's job applications.

Furthermore, the dung fly studies provided for the time unusual quantitative evidence for male-male competition. Parker used optimality models to predict the copulation time—should a male spend the 100 minutes needed mating with one female to fertilize all her eggs, or shorter so that he could also fertilize other females' eggs? Parker hypothesized that males would optimize their time mating with each female in order to fertilize the total maximum amount of eggs. He used the sterile male technique (see Chap. 5) to work out the proportion of eggs fertilized by the last male to mate,[15] and determined experimentally the proportion of eggs a male fertilized at different copulation times. The observed average copulation time is 38 minutes, close to the model's optimum value,[16] thus supporting the prediction that males optimize their time mating with each female. Thus, these studies corroborated the notion that dung flies are strategists and behave according to what gives optimal number of fertilizations, but they were largely focused on male strategies.

---

[13] Simmons, 2001: ix.

[14] Geoff Parker interview 2017.

[15] Parker, 2010b: 437.

[16] Seeley & Sherman, 2021.

## Female Perspectives

Geoff Parker recalls that as honor students, he and his fellow zoology student Robin Baker (who later pioneered in the controversial area of human sperm competition) struggled to understand natural selection in terms of individual interests. Parker says he was thus interested in both males and females, but that he focused on males because when looking at a dung pad it is obvious that yellow dung fly males are fighting furiously to grab females. He also studied females, described their behavior in detail and wrote a paper focusing on female behavior: "The reproductive behaviour and the nature of sexual selection in *Scatophaga stercoraria L.* (Diptera: Scatophagidae) V. The female's behaviour at the oviposition site" (1970b). It was the last paper in a series of five that came out of his dissertation work. The paper focuses on investigating changes in female attractiveness (whether males try to mate with them) and receptivity (whether they allow males to mate with them) as well as factors influencing the male passive phase (when mating is over but the male hangs on to the female's back to prevent other males from mating with her), through the study of female behavior at the dung and its effects on males. Note that both "attractiveness" and "receptivity" are passive formulations of female behavior from a male perspective. Nonetheless, Parker discussed the benefits of male mate guarding after copulation for both males and females, it is beneficial to males for increasing paternity, and for females to prevent wasting time in remating or rejecting another male. Parker found that females need both mature eggs and the presence of dung in order to allow the male to begin mate guarding, or females avoid it through showing swaying responses during mating. The optimal female strategy is to "be receptive" as long as she has eggs left, because rejecting males causes time waste. According to Parker, female dung flies are in a difficult position—there are always males around fighting and desperately trying to mate, and a female easily gets enough sperm to fertilize her eggs through one single mating—so it is debatable whether it is in her best interest to mate again. However, in his study, he also describes the intricate details of female behavior (cited above) when displaying in front of a male, which I would describe as an active female strategy even if it only happens on rare occasions, although Parker seems to have theorized females as passive. Parker recalls that he felt slightly frustrated at being accused of only seeing the male perspective, because with his focus on selection on the individual level he had also tried to work on females, but he says his paper on female behavior didn't attract much interest.

In "Sperm competition and its evolutionary consequences in the insects" (1970) Parker reviewed earlier studies in the light of the sperm competition hypothesis.[17] The review covered references from the broad entomological literature on evidence of the prerequisite for sperm competition, female multiple mating, in natural history observations, population genetics and applied research (see Chap. 5). The review profoundly focused on males and male interests, it concerned mainly the male mechanisms that could increase the chances of their sperm fertilizing most eggs.[18] Yet, near the end of the paper, Parker pointed out that "The female cannot be regarded as an inert environment in and around which this form of adaptation evolves"[19] and developed an argument around male benefits of mating plugs and expectations that females will evolve modifications to prevent or reduce disadvantages of plugs. I also found a citation that I thought might hint at ideas about female choice and female influences on fertilization: "That the female may affect sperm viability has been shown by Oliver and Anderson (1945),"[20] but Parker does not remember that it made him think about female influences. Here's another example of production of ignorance, there is evidence of females affecting sperm viability, Parker even cites it, but he does not incorporate that when formulating his ideas about (male) post-copulatory sexual selection. He mentions female sperm storage organs and multiple mating as preconditions for frequent sperm competition in insects, but otherwise focus on male adaptations—sperm displacement, mating plugs, prolonged copulation, mate guarding, territoriality and mechanisms to avoid mate take-overs. Parker explains that "males were easier to study than females, it was just so much easier to look at their behaviour and what they did".

---

[17] Parker, 1970b.

[18] Ibid.: 528.

[19] Ibid.: 559.

[20] Ibid.: 530.

## Being in a Male Phenotype

Geoff Parker situates his starting point in his male embodiment, "Now there is the issue of course of whether being in a male phenotype has some impact on that." He explains further, "If you study sexual selection [in contrast to nuclear physics], unfortunately I have to admit that your own phenotype is likely to have some impact on the nature of how you think about it. And I regret that that is so. But it's a fact of life, I think. I mean, you exist in a female phenotype and I'm sure you have, rightly so, a female perspective, and I in a male phenotype and vice versa. That doesn't mean though, that we should close our minds to the opponents… the opposing point of view. We should indeed try to seek the truth, and that is often to see both points of view as best we can." In addition, he talks approvingly about the influx of women in science leading to a more balanced perspective. Although I agree with Parker that the experience of being male or female has some impact on science, I prefer to think about it in terms of situated knowledges, that our embodied and social experiences influence our knowledge-making. The notion that men and women bring their own biases that cancel out each other has been expressed by some feminist primatologists as well.[21] However, it seems to imply an essentialist notion of women and men. Furthermore, science theoretician Donna Haraway criticizes this perspective for adhering to the traditional philosophy of science as objective.[22] In contrast to traditional philosophy of science, Haraway's situated knowledges provides an epistemology (theory of knowledge) that takes into account the complexities of researchers' experiences, theoretical perspectives and societal influences on the process of knowledge production (see Chap. 1).

Sexual selection history also shows that "female perspectives"[23] are not restricted to one sex, for example, Randy Thornhill and William Eberhard (later in this chapter) are proponents of ideas about female agency through cryptic female choice.

---

[21] For example Thelma Rowell and Jane Lancaster in *Female Primates: Studies by Women Primatologists* edited by Meredith Small, 1984, see analysis in Haraway, 1989: 290ff.

[22] Haraway, 1989: 290ff.

[23] "Female perspectives" is an often-used term in evolutionary biology meaning to consider the world from a female point of view, as opposed to an androcentric perspective.

Parker argues that you have to start somewhere, and from his perspective, the investigations he thought of were mainly related to males—they were easier to observe and perhaps these questions came more easily to him. Since the theory of sexual selection is formulated with focus on how some males gain higher reproductive success than others (see Chap. 3), it would maybe have been surprising if Parker's work had not been focused on males.

## The Reception of Sperm Competition

Apart from some criticism from some ethologists and ecologists who supported the group selectionist's perspective, Geoff Parker experienced very little resistance to his idea about sperm competition. Population geneticists and evolutionary biologists were not opposed to the idea. However, as with many new ideas, it wasn't noticed much initially, the interest increased after a sperm competition conference in 1980 organized by entomologist Robert Smith in Tucson, USA, followed by an edited book[24] and then increased considerably in the 1990s (Fig. 6.2).

## Response to Cryptic Female Choice

I asked Geoff Parker about his response to the idea of cryptic female choice and he admits that he initially was skeptical. In Parker's lab in the early 1990s, the postdocs, among them entomologist Leigh Simmons, used to joke about his "slight apprehension of female choice". The problem as he then perceived it concerned additive genetic variance, he explains that "whichever sex chooses a trait, you'd expect that trait to fix (become invariant) in the other sex and then it's no basis for a choice". In other words, the choice for a trait would select that trait to spread in the population and soon lead to the whole population having the same trait, then mate choice would not matter anymore. He says it seemed perfectly

[24] Smith, 1984.

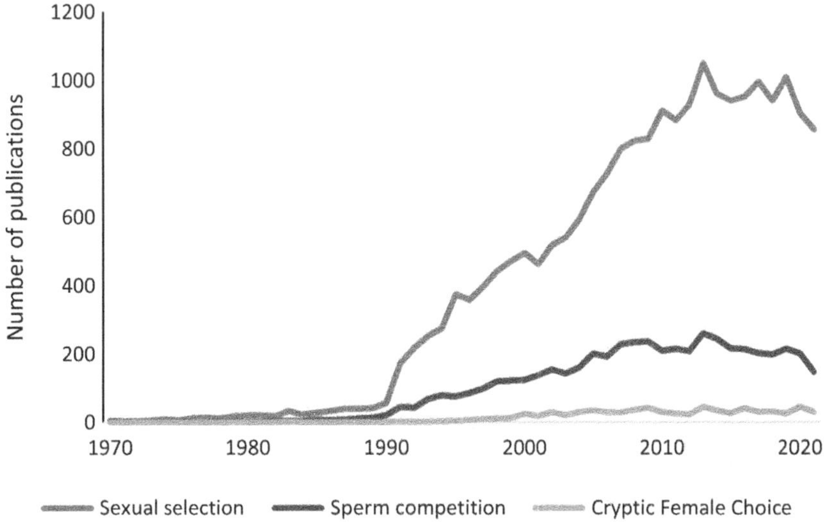


**Fig. 6.2** Number of publications on sexual selection compared to publications on sperm competition and cryptic female choice respectively over time. Sperm competition studies by far outnumber cryptic female choice studies. A number of publications were retrieved from Web of Science searches in "all fields" for "sexual selection", "sperm competition" and "cryptic female choice" (June 6, 2022). Reprinted from Ah-King (2022) under the Creative Commons CC BY 4.0 license

logical and sensible and was therefore a fairly powerful objection at the time. Now we know that it doesn't hold up, he adds, because there are so many genes and random mutations that there will always be a lag and a certain amount of genetic variance.

Parker says he remained skeptical of cryptic female choice until the mid to late 1990s, and was then influenced by discussions with his research group, particularly with Leigh Simmons, and came to a gradual acceptance. Parker declares he has come around because now there is evidence of cryptic female choice to a certain extent—although not as much evidence as for male-male competition. He adds, "I never doubted that females exercise choice in the human species, so it seems entirely feasible that in other animals too, that should be applied."

## Parker's Situated Knowledges

Parker's situated knowledges are strongly influenced by his study species and sexual selection theory. Parker chose to study dung flies, a system that is dominated by male competition, and because the last male to mate fertilizes 80 to 85 percent of the eggs (because their sperm displace earlier males' sperm), there is little room for female influence over fertilization in this species (see the section on Leigh Simmons below).

Thus, Parker's situated knowledges, with yellow dung flies as models, sexual selection theory, his primarily male perspective and theoretical models, have been and still are very influential in contemporary sexual selection.

## Randy Thornhill: Coining Cryptic Female Choice and a Career Ripe with Controversies

Entomologist Randy Thornhill, Professor Emeritus at the University of New Mexico with a PhD in Zoology from the University of Michigan (1974), coined the term cryptic female choice in a paper on the scorpion fly, *Harpobittacus nigriceps*, in 1983.[25] I was curious to meet Thornhill, since he is a controversial figure in the scientific community. Especially controversial was his book *A Natural History of Rape*[26] published in 2000, together with anthropologist Craig Palmer, in which they compare "rape" in insects and rape in humans, and argue that rape is a sexually motivated adaptive strategy or a byproduct of other adaptive traits (sexual desire or aggressiveness). In addition, they suggest a number of strategies to reduce rape in society, such as to inform young men that they are evolutionarily prone to rape and warn young women to check how they dress. The book ignited an outrage and received massive criticism from feminists, anthropologists, evolutionary biologists, primatologists, sociologists and

---

[25] Thornhill, 1983.

[26] Thornhill & Palmer, 2000; Criticism against the use of the term "rape" in animal behavior has been around since the 1980s, see, for example, Gowaty, 1982.

psychologists.[27] As I approached Thornhill's office in Albuquerque to conduct the interview, I was intrigued to know how he came to both suggest that rape is an evolved strategy among humans and envision the female-centered hypothesis of cryptic female choice—it seemed like a feminist paradox.

Thornhill traces his research interests to his childhood when he collected insects and his mother early on introduced him to Charles Darwin's writings.[28] He says that his research is motivated by "a fundamental interest in the female phenomena". Since he has spent so much of his research working on females, people have asked him whether he had "odd or interesting relationships with women growing up". Yes, he affirms, he was raised in the old south by "a bunch of women—my mother, grandmothers, aunts" and as he was the only child, all these women totally invested in him. "And then, of course when I hit puberty I took a different interest in the female phenomena" he adds laughing, here alluding to (hetero-) sexuality (Fig. 6.3).

## Developing the Idea of Crypic Female Choice

Randy Thornhill recounts how the idea of cryptic female choice emerged from his empirical work on hanging flies and scorpion flies. His first field study on the hanging fly *Hylobittacus apicalis*[29] was published in 1976 and showed that females were choosing males on the basis of nuptial gift size. (Nuptial gift is an anthropomorphic metaphor for a nutritional offering provided at mating, in hanging flies, it is a prey item.) "I followed these females, I saw them turn down males" he recalls, then he continued working with scorpion flies that showed very conspicuous female choice related to nuptial gift size. Starving females turned down males with small prey at high population densities, but at lower densities, females were less choosy. They would mate with a male with small prey,

[27] See Travis, 2003. The controversy around the book is also part of a bigger controversy around evolutionary psychology that concerns its basic assumptions and methodology, see, for example, Rose & Rose, 2000.

[28] Thornhill, 2015: vii.

[29] The species name *Bittacus apicalis* has changed to *Hylobittacus apicalis* since 1976.

**Fig. 6.3** Randy Thornhill in his office in Albuquerque, University of New Mexico, USA

but then stop mating after five minutes. This observation puzzled Thornhill, "the prey is big enough for her to feed for 15 or 8 minutes, why wouldn't she? So, she's making this very sophisticated decision: five minutes! And I showed that's when the sperm from the male begins to flow into the female's body." Thus, when the prey was small she obtained food but avoided the male's sperm. In this case, female choice selects for males that can hunt big prey, but also keep that prey in competition with other males trying to steal it. Females can also make the decision not to lay eggs until they mate with a male with large prey. This optional reproduction strategy occurs both in hanging flies and scorpion flies. Here, Thornhill's development of a female-centered idea emerged directly from his empirical observations of his specific study species, thus it is another example of the influence of study species on researchers' situated knowledges.

## Reception of Thornhill's Research

Randy Thornhill recounts a memory of the early response to his research. At his first presentation at a conference, an international entomology conference in 1978, he presented his first study on female choice in hanging flies. "Those people thought I was crazy. It was at an insect conference and so I gave the talk and showed them evidence of female choice for nuptial gift size." The response by the audience was to riducule the study, "people were making jokes about it [...] this guy in the audience would say something about 'these females are prostitutes' and then those other guys would have another joke and everybody [was] laughing. I felt like maybe I should just drop out and go to medical school." But he didn't drop out and has over time become used to controversies—"all this stuff I've done has been controversial, at some level". "People like me, we just blast away", criticism seems to be like water off a duck's back to him, he says he lets other people take care of the reactions. Ignoring the criticism, he asserts, is "associated with developing a thick skin, being tough". Although the controversies have also led to serious threats, after publication of *The Natural History of Rape*, Thornhill received death threats and got campus police escort to and from classes.

Indeed, he has ventured into controversial areas of research, 25 years ago he turned to human sexuality. Thornhill urges me to write more about human sexuality, "that would sell!", gives me a private lecture on "women's dual sexuality" (women, according to him, choose provisioning mates for long-term partners and then around ovulation copulates with extra-pair mates of high genetic quality) and donates his book on human female sexuality.

## Coining Cryptic Female Choice

The term cryptic female choice was coined (meaning female choice during or after mating) in Randy Thornhill's 1983 article about scorpion flies. However, the conception of active females is prominent already in his 1976 paper on female choice in hanging flies, in which he investigates just how female choice works—which size and kind of prey they prefer, how long they copulate with a male. When a female terminated a copulation with one male, Thornhill followed her without taking his eyes off her until she mated with another male.[30] Thus, his detailed observations led him to conclude that females had agency in determining the length of copulations and whether to copulate at all.

Working on scorpion flies added other dimensions of female choice, Thornhill recounts, "I had like all kind of alternative ideas about things that had traditionally been interpreted as sperm competition in that paper[31] [...] there's a lot of forced copulation in the scorpionfly system, male forced copulation and females try to get away from these males. Males have an organ that holds the female's wing that functions in rape, it's a rape organ." If one "render[s] that organ nonfunctional, males can't rape", asserts Thornhill. When a female has been copulated by force, she will copulate quicker with another male (return to sexual receptivity) than if she has been given a nuptial gift in a consensual mating. "So, basically selecting against the rapist, there's a kind of indirect female choice there." Thus, he describes the dynamics of sexual selection, both the male strategies and the female counter-strategies.

In a book chapter from 2015, Thornhill delineates the history of cryptic female choice: "The thinking that lead to original CFC [cryptic female choice] theory arose independently from a number of researchers."[32] He mentions entomologist James Lloyd (known for his work on fireflies) and entomologist John Sivinski at the University of Florida (see also Chap. 5). At the symposium on "the Sociobiology of sex"—in which insect behavior and mating systems were discussed in the light of the new

[30] Thornhill, 1976: 543.

[31] Thornhill, 1980.

[32] Thornhill, 2015: ix.

sociobiological framework (studying social behavior from an evolutionary perspective)—James Lloyd discussed post-copulatory sexual selection in both males and females already in 1978.[33] According to Thornhill, the entomologists did not show any resistance. In the paper following the symposium, Lloyd wrote "After sperm has been placed in a female, she should manipulate it: store, transfer (from chamber to chamber), use, eat or dissolve it, as she makes additional observations of males."[34] He further points out that "it is possible that some reported examples of sperm competition are actually cases of sperm manipulation."[35] When I asked Thornhill whether his, Lloyd's and Sivinski's were independent origins of the idea, he answers they were not independent but "kind of collective". Thornhill says he "was very hot on the subtle female choice" and while he was working in Florida for a year he interacted a lot with James Lloyd, and John Sivinski was Thornhill's graduate student. The three of them were collaborating and discussing these issues frequently.

Another entomologist, John Alcock, was also very supportive of the idea of cryptic female choice, he recognized that the issue had been ignored and needed to be put into the mainstream. Thus, Thornhill and Alcock's *Evolution of Insect mating systems* (1983) was early to include female perspectives on post-copulatory mating, and became influential in entomology as well as beyond. The book is described, in the preface to the 2014 follow-up book with the same title, as canonical for the field of behavioral ecology by explaining "the complex and often strange sex lives of insects" in an evolutionary framework.[36] Thus, insect research was the subfield in which female perspectives on post-copulatory mating developed, after the male perspectives on sperm competition.

## Development of Cryptic Female Choice Research

Randy Thornhill's cryptic female choice paper from 1983 has few citations in the subsequent years after its publication. "There really was not

[33] Thereafter published in 1979 in the Florida Entomologist.

[34] Lloyd, 1979: 22.

[35] Ibid.: 22.

[36] Shuker & Simmons, 2014: vii.

much action for cryptic female choice until Bill [William] Eberhard's book *Female Control* came out in 1996 and it really took off in terms of citations" he recalls (see also Fig. 6.2). Eberhard's book showed that cryptic female choice was a larger phenomenon than just the observations in hanging flies and scorpion flies. Still, Thornhill has been influential in entomology; several renowned entomologists such as Darryl Gwynne (see Chap. 5), Marlene Zuk (expert on sexual selection and parasites) and Göran Arnqvist (entomologist and sexual conflict researcher) were postdocs in his lab; he also told ornithologist Tim Birkhead about his observations on female sperm rejection in red jungle fowl. Birkhead later pioneered cryptic female choice studies in birds with a study of this species (see section on cryptic female choice in birds below).

According to Thornhill, the criticism toward cryptic female choice was that "it ignores the important role of sperm competition and the male control of all that sperm". He ascertains that Geoff Parker's insight about sperm competition was very important and has resulted in a tremendous amount of research, "but when a female's inseminated by multiple males you don't expect them to be passive about it!" Cryptic female choice provides a supplementary understanding of post-copulatory sexual selection. He adds that all female choice in plants is cryptic female choice, because they don't have the pre-mating context (when pollen from several plants lands on the pistil they grow pollen tubes toward the ovules and plants discriminate between pollen grains[37])—but "Botanists have not taken that up very much."[38]

Thornhill recalls that Geoff Parker and William Eberhard had a big argument at a conference in London due to their different perspectives—in which Parker was backed up by Tim Birkhead and other supporters. Hence, there was an initial controversy between researchers of sperm competition versus cryptic female choice. By presenting an alternative hypothesis, Thornhill also criticized sperm competition theory

[37] Moore & Pannell, 2011.

[38] Despite Thornhill's assertion here, there is interesting cryptic female choice work going on in plants, for example by Leonie Moyle and colleagues at Indiana University, USA. Furthermore, Åsa Lankinen and colleagues at the University of Agricultural Sciences in Alnarp, Sweden, have explored the implications of sexual selection in plants broadly, and specifically investigated male-male competition.

for being male-biased, arguing that "everything males and females do can't be just sperm competition". In the preface of Robert L. Smith's (1984) edited volume *Sperm competition and the evolution of animal mating systems*, Thornhill's chapter is described as "a caveat": "Randy Thornhill discovers alternative contexts for characters initially believed to have evolved by sperm competition; his contribution cautions the reader and instructs the prospective investigator in the field."[39] Accordingly, Thornhill's chapter is entitled "Alternative hypotheses for traits believed to have evolved by sperm competition". There he describes how he first interpreted the male scorpion fly dorsal clamp organ (used to hold females during mating) as evolved for reduction of sperm competition, but later on found that it functions in forced copulations.[40] In this account, females are active—they avoid males without nuptial gifts, they manage to break free from males that attempt to copulate by force in 85 percent of the cases, return to sexual receptivity quicker and "somehow cause low insemination rate" when forced.[41] Interestingly, in this 1984 chapter, Thornhill does not use the term "rape" (which he did in 1980) but instead "forced copulation", perhaps in the wake of the criticism against its use in animal behavior. But in *A Natural history of rape* (2000), as well as in my interview, he uses the term "rape" again. Thus, his shifts in terminology do not seem to emerge from a gender-reflective perspective on language use.

During the interview, Thornhill relates to our mutual interest in what he calls the "female phenomenon" and criticizes what he calls "gender feminists". As we discussed his research, I realized that from Thornhill's perspective there is no feminist paradox. He simply focuses on biological explanations, he found coercive matings among the scorpion flies, and also that females had counter-strategies, so from a sociobiological perspective, the explanations about male and female behavior are just two sides of the same coin—feminist perspectives are not needed to reach this conclusion. Thornhill's situated knowledges about sex and sexual

---

[39] Smith, 1984: xviii.

[40] Thornhill, 1984: 152.

[41] Ibid.: 163.

selection are influenced by his study species, his meticulous studies of insect matings and his theoretical perspectives and his work have thus highlighted female agency in sexual selection.

## William Eberhard: Developing the Big Picture of Cryptic Female Choice

William Eberhard is an American tropical biologist at the Smithsonian Tropical Research Institute in Panama and renowned for his studies of animal genitals and cryptic female choice. He received his PhD in Biology on spiders from Harvard University in 1969. A quote on his homepage reads: "Females are often mistakenly thought to play relatively passive roles in copulation and reproduction, but their cooperation is often crucial in determining whether or not a copulation results in reproduction."[42] He actively challenges prevailing assumptions in sexual selection through his development of a comprehensive understanding of females' multiple abilities to shape the outcome of matings.

### Interest in the Diversity of Animal Genitals

In our correspondence over e-mail, William Eberhard explains that his interest in cryptic female choice grew out of an interest in the evolution of animal genitals. As an undergraduate biology student in the 1960s, he earned extra money as an assistant at Harvard University's Museum of Comparative Zoology by organizing unsorted spiders.[43] Using spider identification guides, he typed spider species and started to wonder why spiders are so often identified by genitals—among spiders as well as many insects, the genitals are often very different even between closely related species. The museum seniors answered "that's just the way it is" and told him that the genetic differences probably accidentally affect the shape of the genitals which is useful for taxonomists (whose work is to classify spe-

[42] https://stri.si.edu/scientist/william-eberhard.

[43] Schilthuizen, 2014: 3.

cies), but probably biologically meaningless.[44] He wasn't convinced with this explanation, but also not in a position to challenge it, so the question was left dormant as his career developed successfully in another direction, namely tropical biology.[45] It wasn't until entomologist Jonathan Waage's 1979 pioneering paper on the sperm-scooping penis (used to first scoop out previous males' sperm and then deposit own sperm) of the damselfly *Calopteryx maculata*, that Eberhard picked up the question of evolution of genitals again, thinking that perhaps the divergent genitals were different kinds of sperm-scoops?[46] Thereafter he spent time immersed in the library to review studies of animal genitals in the light of sexual selection, which led to the publication of the book *Sexual Selection and Animal Genitalia* (1985). In this book, Eberhard reviews the diversity of male genitals from nematodes to snakes and rodents, showing that animal genitals evolve relatively faster than other body parts. More precisely, it is in internally fertilizing species that male genitals are diverse and complex; in contrast, male genitals in species with external fertilization are consistently simple and uniform. He therefore suggests that male genitals in internally fertilizing species are under intense sexual selection. He reviews several hypotheses and suggests a new one, namely that female choice of internal "courtship devices" led to genital diversity and concludes that the latter hypothesis is the only one that can explain the majority of cases (although other hypotheses may be true in some cases). This hypothesis is in essence an argument for cryptic female choice.

In *Sexual Selection and Animal Genitalia*, Eberhard also lists the difficulties with his hypothesis, for example that it can only work if females generally mate with multiple mates during their reproductive lifetimes.[47] The data on remating was scarce at the time, but among butterflies, which provided the most complete information (because it is possible to count the number of spermatophores in the female reproductive tract), remating was known to be very widespread. He notes that genital morphology can change rapidly in a runaway process without affecting survival, in contrast

[44] Ibid.: 3.

[45] Ibid.: 3.

[46] Ibid.: 4.

[47] Eberhard, 1985: 184.

to runaway processes of pre-mating choice for elaborate plumages. (As described in Chap. 3, Ronald Fisher's runaway hypothesis predicts that a female preference for a trait can over time co-evolve with the preferred male trait and lead to exaggerated male traits that "run-away" from control by natural selection, and thus become costly in terms of survival.) In the acknowledgments of the book, Eberhard gives thanks to a number of women scientists (eight out of 66 named scientists) and one of them is his wife, the renowned evolutionary biologist Mary Jane West Eberhard. Compared to ornithologist Malte Andersson's lack of women colleagues in the acknowledgments of *Sexual Selection* published in 1994 (discussed in Chap. 3) almost a decade later, Eberhard seems to have more women in his intellectual network. Women scientists did occur in evolutionary biology at this time, whether the difference between the networks is a matter of differences in Andersson's subfield ornithology and Eberhard's entomology, or a difference between personal networks, is difficult for me to tell. However, as a parallel to the influx of women in primatology simultaneously with the development of feminist perspectives (see Chap. 2), Eberhard with more women in his network, also shows a more gender-reflective approach calling out gender bias in science.

## Ignorance of Cryptic Female Choice

William Eberhard writes that his goal with *Sexual Selection and Animal Genitalia* was to bring genital evolution into the mainstream of evolutionary theory,[48] but he says that it was mainly received as of narrow interest for those doing research on animal genitals, that is essentially taxonomists. He states that the book was largely ignored by researchers in sexual selection. Eberhard especially remembers two papers on sexual selection ignoring cryptic female choice, that were published in 1991 in *Nature*[49] and *Trends in Ecology and Evolution*[50] respectively, written by three prominent evolutionary biologists. Reading these papers became Eberhard's incentive for writing *Female Control: Sexual Selection by Cryptic*

[48] Ibid.: 188.

[49] Kirkpatrick & Ryan, 1991.

[50] Smith, 1991.

*Female Choice*, a book that was published in 1996. In *Female Control*, he considered multiple theoretical possibilities for females to exert choice in post-copulatory situations and collected and reviewed the evidence for it. He defined cryptic female choice as: "Sexual selection by cryptic female choice can result from a female-controlled process or structure that selectively favors paternity by conspecific males with a particular trait over that of others that lack the trait when the female has copulated with both types."[51] He started out by arguing that cryptic female choice has been largely ignored, "despite the much greater opportunities that females have to influence events."[52] He called attention to the biases that the previous research has been pervaded with: the assumption that males and females cooperate during mating for the good of the species, and the assumption that females are passive. He pointed out that, although there are exceptions, there is a general male-centered tradition in biology and "recent studies routinely emphasize the male side of male-female sexual interactions."[53]

Eberhard also reinterpreted some earlier findings, and pointed to possible alternative interpretations of experiments of male sperm precedence (variability was overlooked and could be due to cryptic female choice), of restricted conditions in the same experiments (do they correspond to conditions in nature?), of the effects of male seminal products (variability is overlooked and could be an effect of cryptic female choice) and of unnecessary strict interpretations of female benefits in a species (females could benefit in multiple ways). Thus, he challenged established interpretations in sperm competition and physiological research in sexual selection.

In a characteristic tone, he emphasizes empirical studies: "As an empirical biologist, however, I have developed antibodies to presentations of intellectually titillating theoretical possibilities made without any serious attempt to check provocative ideas against the hard realities of nature. [...] I have thus tried hard to impose a further bias and discuss only feasible types of interactions that flow from 'reasonable' interpretations of

[51] Eberhard, 1996: 7.

[52] Ibid.: 5.

[53] Ibid.: 34f.

hard data from real animals."[54] The book provides an encyclopedic list of possibilities of cryptic female choice, mainly in insects and mammals, but with some examples from nematodes, birds and snakes.

He suggests more than twenty different mechanisms by which females may control which sperm fertilize their eggs. These include morphological (lead the sperm astray), physiological (destroy the sperm) and behavioral (disrupt copulation) ways to either obstruct or facilitate fertilization. Thus, he pulls together a multitude of dispersed evidence of potential cryptic female choice from morphology, reproductive physiology and behavioral studies, and suggests that cryptic female choice is widespread.

Eberhard argues for acknowledging that females are active, and cites primatologist Sarah Blaffer Hrdy's *The Woman that Never Evolved* (1981) as well as science writer Mary Batten's "myth of the passive female" in *Sexual Strategies: How Females Choose Their Mates* (1992). This myth "needs to be discharged in studies of behavioral ecology, physiology and morphology of reproduction", concludes Eberhard and emphasizes that the idea of cryptic female choice focuses on how females set "the rules of the game."[55] Subsequently, he repeatedly urges for an increased understanding of female agency, for example, in 1990, he comments on a review paper on mechanisms of sperm competition with the critical reply "Inadvertent machismo?" in which he points to several instances when the authors overly focused on male perspectives and assumed females to be passive.

In our correspondence, Eberhard underlines the importance of distinguishing between "ignoring an argument (or dismissing it with the single word "controversial", ignoring all the data and arguments that might be relevant), as opposed to actually disproving it with data or logical arguments". He finds it frustrating that cryptic female choice has often been ignored in this way and continues to be so, which I get back to in the next section on the controversy over cryptic female choice.

Eberhard's particular situated knowledges are based in his empirical studies of insects and spiders, his insights of male bias in previous interpretations and theoretical perspectives encompassing a multitude of ways

---

[54] Ibid.: 21.

[55] Ibid.: 420.

in which females may exert cryptic choice. Thus, his work has been pivotal for expanding sexual selection to include active females in post-copulatory sexual selection.

## Controversy Over Cryptic Female Choice

As we have seen, the cryptic female choice idea was not much cited during the years following Randy Thornhill's 1983 paper, citations started increasing after the publication of William Eberhard's book *Female control* (1996), which also stirred up a controversy around the issue of cryptic female choice. Herpetologist Mike Ryan describes the reactions among biologists as "Well, it seems logical, is it biological? Sure it seems like it could happen, but is it really happening? And how important is it?"[56]

Geoff Parker recounts that initially the two sides in the controversy, sperm competition and cryptic female choice, were polarized as if you had to choose either side and either believe in sperm competition or cryptic female choice. For a time then, the different sides seemed diametrically opposed, and Parker asserts that like in other controversies, here too "people defend what they see as their thing, and the other thing as in opposition to it."[57]

### Initial Response: Reviews of Female Control

The scientific community's early responses can be seen in the reviews of *Female Control*, which praise the gigantic effort to summarize all the different ways in which cryptic female choice may occur. For example, entomologist Douglas Tallamy called it a "monumental synthesis of a disparate literature" that "exposes a blatant and embarrassing male bias in much of the historical theory and research on reproductive behavior" and was already then appointed a central reference.[58] Geneticist reviewers Brett Holland and William Rice also acknowledge the "Herculean achieve-

[56] Mike Ryan interview 2017.

[57] Geoff Parker interview 2017.

[58] Tallamy, 1997: 485–486.

ment" to survey diverse animal groups and to show that female anatomy and physiology are significant for male reproductive success. They were convinced that the female side of post-copulatory sexual selection has been overlooked, but they don't agree to all of Eberhard's interpretations. They think that Eberhard overemphasizes female control, because neither sex is expected to ever be completely "in control."[59] Even though a female may disrupt a mating and thereby "control" the sperm transfer, male seminal fluids have been shown to influence female remating rate and mortality. Furthermore, they are not convinced by Eberhard's interpretation that cryptic female choice causes rapid evolution of male genitals, instead they emphasize sexually antagonistic co-evolution at the genetic level as an explanation, that is evolution of traits that are beneficial for males but disadvantageous for females (or vice versa). Thus, their review shows the emerging controversy between the competing hypotheses of cryptic female choice and sexually antagonistic co-evolution (sexual conflict).

## Recognition and Skepticism of Cryptic Female Choice

Neither entomologists Darryl Gwynne nor Randy Thornhill remembers any resistance toward cryptic female choice among their entomology colleagues at the Sociobiology of Sex symposium in Gainsville, Florida. Gwynne says that once the idea was out, the reaction was "wow, this is revolutionary! What a way to think about this, this is fascinating, let's go to work on this!" In contrast, William Eberhard describes instances of "quite energetic resistance" among some entomologists and ornithologists. In 1997, entomologist Richard Alexander and colleagues argue against cryptic female choice as the explanation for the evolution of elaborate male genitals and argue for sexual conflict as the typical explanation.[60] Here the controversy between cryptic female choice and sexual conflict is manifested again.

---

[59] Holland & Rice, 1997: 323.

[60] Alexander et al., 1997.

In writing about the history of sexual selection, ornithologist Tim Birkhead has described that cryptic female choice was cast with suspicion when it was suggested.[61] Several authors of chapters in *Sperm Competition and Sexual Selection* edited by Birkhead and ornithologist Anders Pape Møller in 1998, express skepticism toward cryptic female choice. In the preface, Birkhead and Møller describe William Eberhard's chapter as dealing "with a rapidly expanding and controversial area, the role of females in sperm competition". Geoff Parker, in his chapter, remarked that "although there is some evidence for sperm preference by females (e.g. Eberhard, 1996), much of it is flawed, and anything that hints of female control tends to be highlighted enthusiastically."[62] Thereafter Parker mentioned some plausible mechanisms, but was clearly skeptic and dismissive toward cryptic female choice. In the chapter on birds, Birkhead discussed possibilities of cryptic female choice, he argued that the earlier mentioned spotted sandpiper outcome was more probably due to sperm competition (rather than female influence, in contrast to Marcy Lawton and colleagues' gender analysis, see Chap. 4), but also refers to females' possibilities of ejecting sperm and some other mechanisms to bias paternity.[63] Thus, the skepticism came both from entomologists and ornithologists, both from sperm competition researchers and those emphasizing sexual conflict.

## Eberhard's Response to the Skepticism

William Eberhard responded to several authors' skepticism in his own chapter in Tim Birkhead and Anders Pape Møller's 1998 volume, criticizing that Geoff Parker's statement "much of it is flawed" without specific examples does not give any possibility of responding to the critique. Such statements served to cultivate ignorance about cryptic female choice. Eberhard also pointed out that Parker fails to take into account that females set the rules of the game.[64] Eberhard responds to Birkhead and

[61] Birkhead, 1996, 2000a.

[62] Parker, 1998: 47.

[63] Birkhead, 1998a: 607ff.

[64] Eberhard, 1998: 112.

Møller's conclusion that cryptic female choice remains controversial, arguing that although there are few studies on cryptic female choice, the hypothesis has survived several tests without being modified or rejected, and the rapid evolution of male genitals is hard to explain otherwise.[65]

## Debate About Female Sperm Choice

In the wake of the publication of *Female Control* and *Sperm Competition and Sexual Selection*, Tim Birkhead started a debate with his review "Cryptic female choice: criteria for establishing female sperm choice" in 1998. He deemed the evidence for cryptic female choice to be overstated, although he acknowledged the evidence for differential abortion of embryos and differential investment in offspring, he considered the most controversial issue to be whether females choose sperm after copulation. Sperm choice he defined as "The simultaneous recognition of and discrimination between sperm of different males, either on the basis of the males' phenotype or that of their sperm constitutes sperm choice,"[66] thus a much narrower subset of what is included in Eberhard's broader definition of cryptic female choice. Partly, Birkhead writes, female sperm choice is disputed because if it occurs, it would happen at the same time as sperm competition and therefore it would be extremely difficult to distinguish between these processes.[67] Birkhead then set up criteria for determining whether female sperm choice occurs. In the ensuing discussion, it becomes clear that Birkhead and Eberhard continue to disagree over what constitutes evidence for cryptic female choice and female sperm choice. For example, while Birkhead claims that one needs to control the effects of sperm competition in order to demonstrate female sperm choice, Eberhard argues that for many of the female processes influencing paternity it is not necessary to exclude sperm competition in order to find evidence of cryptic female choice, because they are not likely to be confused with sperm competition (for example sperm dumping and

[65] Ibid.: 113.

[66] Birkhead, 1998b: 1213.

[67] Ibid.: 1213.

interruption of copulation[68]). Partly the disagreement seems to be an effect of their narrow versus broad definitions, but it is also Birkhead's focus on trying to find an experimental protocol for examining the most contested issue of cryptic female choice (and in that emphasizing sperm competition and sperm competition methodology) versus Eberhard's defense of the long-overlooked issue of female influence in postcopulatory sexual selection.

## Problematizing the Dichotomy Between Sperm Competition and Cryptic Female Choice

One interesting point that William Eberhard makes in this debate is that describing sperm competition and cryptic female choice as a dichotomy is misleading. He argues that Birkhead's either-or-argument implies that "male effects can occur in the absence of female effects, and vice versa. Neither extreme is possible, however; sperm competition sensu strictu is between male products that are inside the female's body."[69] In other words, the outcome of postcopulatory sexual selection will always depend on the interaction between the sperm and the female environment. This perspective, that sperm competition and cryptic female choice represent a false dichotomy, recurs in recent papers.[70]

## Reformulating the Definition of Cryptic Female Choice

In the subsequent debate, reproductive biologists Scott Pitnick and William Brown pointed out that they found the word "female-controlled" in William Eberhard's definition of cryptic female choice problematic, "because it implies that females exercise authority over sperm use."[71] Instead, they suggest a more neutral definition without using the word

68 Eberhard, 2000: 1048.

69 Ibid.: 1048.

70 For example, Lüpold et al., 2020, McDonough-Goldstein et al., 2022.

71 Pitnick & Brown, 2000: 1053.

"control": "nonrandom paternity biases resulting from female morphology, physiology or behavior that occur after coupling."[72] Birkhead approves of this new formulation, because it resolves a possible confusion about whether female passive processes that bias paternity (e.g., due to different dimensions of sperm storage structure) were to be excluded in cryptic female choice with Eberhard's definition—Pitnick and Brown's includes both female active and passive processes.[73] Furthermore, Eberhard too, in a 2015 book chapter, does not use "female-controlled" in his definition:

> CFC, the post-copulatory equivalent of Darwin's female choice, includes paternity biases which result from differences in the expression of female activities (including morphogenesis) that come into play during and following copulation with different males and that favor males which have particular traits over others which lack these traits.[74]

Thus, there is now a consensus on describing cryptic female choice in more neutral terms and to avoid the word "female-controlled".

## Continued Underestimation of Cryptic Female Choice

In 2015, the editors of *Cryptic female choice in arthropods*,[75] scorpion researcher Alfredo Peretti and spider researcher Anita Aisenberg, assert that "though female roles in biasing male paternity after copulation have been proposed for a number of species distributed in many animal groups, cryptic female choice continues to be sometimes underestimated."[76] Furthermore, they note that "during recent years, the concept of sexual conflict has been frequently misused linking it irrevocably with sexual coercion and sexually antagonistic coevolution in opposition to sexual selection by cryptic female choice, without exploring other alternatives."

[72] Ibid.: 1053.

[73] Birkhead, 2000b: 1059.

[74] Eberhard, 2015: 4.

[75] Arthropods are invertebrates with an exoskeleton, including, for example, insects, scorpions, crustaceans and spiders.

[76] Peretti & Aisenberg, 2015: xi.

Thus, relating to sexual conflict as an explanation often overlooks the possibility of cryptic female choice and cultivates ignorance about active females.

Furthermore, there is a bias in research efforts on sperm competition versus cryptic female choice as demonstrated by the differences in the number of publications on these topics shown in Fig. 6.2. Sperm competition studies increase earlier than cryptic female choice and continues to by far outnumber studies of the latter.

## Current State of Cryptic Female Choice in the Scientific Community

As we have seen, the initial controversy over cryptic female choice versus sperm competition has ceased, both ideas are acknowledged, they are now commonly described simply as postcopulatory sexual selection. Eberhard notices that the cryptic female choice idea has historically often been ignored and continues to be overlooked, but he considers the state of research to gradually improve.[77] A recent chapter by evolutionary biologist Rachel Olzer and colleagues on reproductive behavior in insects mentions three examples of cryptic female choice, and then claims that it "largely remains a black box."[78] In a review from 2017, Renée Firman and colleagues summarized the 20-year anniversary of the cryptic female choice field,[79] concluding that there are few clear demonstrations of cryptic female choice. In a response, Eberhard points out that the review was biased toward sperm use tests, vertebrate studies and the fruit fly *Drosophila*, and that it thereby disregards examples of genital evolution and evidence from other arthropods (the taxonomic group including insects, crustaceans, scorpions, millipedes and spiders) in their list of relatively convincing examples.[80] In a response, Firman and colleagues stand by their conclusion that although cryptic female choice has the potential

[77] William Eberhard Personal Communication 2021.

[78] Olzer et al., 2018: 199.

[79] Firman et al., 2017a.

[80] Eberhard, 2017.

to be a general evolutionary phenomenon, the evidence so far has often been indirect and correlational.[81]

Following up on this debate, Eberhard and entomologist Gerlind Lehmann published "Demonstrating sexual selection by cryptic female choice on male genitalia: What is enough?"[82] in 2019, in which they review a list of corroborated predictions of cryptic female choice of male genital evolution—in the well-studied tsetse flies, *Glossina*, and bush-crickets, *Roeseliana*—concluding that these two examples should be included in lists of cases demonstrating cryptic female choice. Furthermore, they emphasize that cryptic female choice as an explanation of genital evolution is extremely difficult to reject, because there are so many ways in which females may respond to male genitals and in many studies, the hypothesis is rejected without testing several possible alternative mechanisms.

Eberhard asserts that convincing tests to differentiate between the hypotheses of cryptic female choice and sexually antagonistic co-evolution have not been possible, that is "direct, accurate empirical measurements in field situations that replicate the conditions under which the traits evolved that measure the crucial difference between cryptic female choice and sexually antagonistic co-evolution—are the female gains with respect to offspring quality greater or less than the female losses with respect to alterations in her behavior and physiology imposed by the male?"[83] Instead, researchers will have to use indirect tests such as exploring whether "female response to experimental lack of stimulation reduces male paternity", "species-specific structures of male genitals are appropriately designed to stimulate the female" and a number of other possibilities listed.[84]

Eberhard comments that conflict-based interpretations often seem to prevail and on a superficial level of analysis male-female interactions, that is "almost trivially true", because male and female interests are seldom

[81] Firman et al., 2017b.

[82] Eberhard & Lehmann, 2019.

[83] William Eberhard Personal Communication 2021.

[84] Ibid.: 2417.

identical.[85] However, in a more narrow analysis of conflict, the controversy between cryptic female choice and sexually antagonistic co-evolution explanations remains: "It is possible that the selection on males to win out over other males has produced profound conflict between the sexes, favoring traits that inflict damage to the female's naturally selected reproductive interests, and that the most important payoff to females from rejecting some males is from avoiding this damage. It is also possible that the most important payoff to the female is obtaining superior survivorship or signaling genes for her offspring and that many cases of apparent male-female conflict are better understood as selective female cooperation with some males but not others."[86] In fact, he adds, different types of selection can act simultaneously on the same trait and different traits in an animal can evolve due to different selection pressures.

Eberhard claims that the prevailing focus on conflict is due to lack of appropriate analyses, and therefore he emphasizes that different hypotheses are likely to be correct in different cases. Although the cryptic female choice hypothesis is accepted among specialists in the field of post-copulatory sexual selection, the more general acceptance among biologists has lagged behind, as exemplified by its treatment in comparison to sperm competition and sexually antagonistic coevolution in a recent evolution textbook.[87]

Furthermore, Eberhard suggests a possible cultural bias toward interpretations of males rather than females as in control. "Another possible problem for general acceptance of CFC [cryptic female choice] may be its emphasis on the female's sometimes potentially invincible powers to control post-copulatory sexual events, in contrast to the perhaps more intuitively appealing views of male control (SC) [sperm competition] or male–female battles over control (SAC) [sexually antagonistic co-evolution]".[88] This is another example of Eberhard's gender-reflective perspective[89] of bias in science, pointing to the interaction of society and science.

---

[85] Eberhard, 2015: 19.

[86] Ibid.: 19.

[87] Ibid.: 20.

[88] Ibid.: 20.

[89] Richardson, 2013: 17.

Thus, the initial controversy over cryptic female choice as an alternative to sperm competition has ceased, since cryptic female choice has become an accepted alternative hypothesis. However, the controversy over cryptic female choice and sexually antagonistic co-evolution interpretations continues.[90]

## Are Females Harder to Study?

To get yet another perspective on the development in the field of sperm competition and cryptic female choice, I consulted evolutionary biologist Caitlin McDonough-Goldstein who is an expert on fruit fly *Drosophila* gene expression in the female reproductive tract. She problematizes the often-repeated notion that the female side is harder to study. She agrees that in some cases it is practically harder to investigate, as with the internal female genitals compared to the external male ones in insects, but given the tepid reception of cryptic female choice in the scientific community as well as the lack of established methodology, there are both theoretical and methodological reasons contributing to making it hard to study the female side. "The greater critique on CFC [cryptic female choice] means that proxy measures such as testis size aren't as readily accepted and/or haven't been established"[91] reasons McDonough-Goldstein. Since species with high degrees of sperm competition have been found to have larger testis sizes, comparing testis size between different species has become a proxy measure for the degree of sexual selection. There is no equivalent proxy measure for cryptic female choice. Thus, both the prevailing androcentric theory as well as (the lack of) established methodology in the field contributes to ignorance production of cryptic female choice.

[90] Ibid.: 19.

[91] Caitlin McDonough-Goldstein Personal Communication 2021.

## Leigh Simmons: Female Choice and Sperm Competition in Insects

Leigh Simmons is a professor at the School of Biological Sciences, Centre for Evolutionary Biology at the University of Western Australia, with a PhD from Nottingham University, UK (1986). He studies sperm competition and genital evolution primarily in insects. He is a prominent researcher in the field and represents a younger generation in relation to Geoff Parker, Randy Thornhill and William Eberhard, it is therefore interesting to follow his approach to the topics of sperm competition and cryptic female choice. I met him at his office in Perth for an interview.

Simmons retells that as a young child he was not allowed to have pets, so instead he started to collect insects in jars in the garden.[92] At the University, it was especially a course in animal behavior, in which Simmons did a laboratory study on dominance hierarchies in crickets, that steered him into studying insects. Randy Thornhill and John Alcock's book *Evolution of Insect Mating Systems* 1983 (which Simmons describes as full of focus on females) steered him toward working with female choice in crickets. This book, which among other things summarized the work from the earlier mentioned Florida entomologists and Geoff Parker's work, was a major impetus for many researchers who started working on female choice. Simmons wrote about cryptic female choice early on in his study on crickets in which the male provides nutritional spermatophores to the females at mating, without using that specific term, he called it "female choice in the field cricket" in 1986.[93] In follow-up articles, he showed how females manipulate insemination by removing spermatophores from some males and not others. In other words, this is an example of cryptic female choice that, like Thornhill's scorpion flies that interrupted mating, is fully observable.

Simmons remembers that as a postdoc with Geoff Parker in Liverpool, they had many discussions and he used to tease Parker for having trouble believing that female choice could occur. As Parker explained, at the time he thought that female choice would soon lead to the fixation of the

[92] Leigh Simmons interview 2018.

[93] Simmons, 1986.

chosen trait, and when there was no more genetic variability left, there would be no basis for choice (but the development of knowledge about genetic variability since then has made this argument obsolete). However, Simmons would point to the evidence that variation is maintained in nature and (jokingly) tell Parker it was his problem as a theoretician to think of how it could be maintained (Fig. 6.4).

**Fig. 6.4** Entomologist Leigh Simmons in his office at University of Western Australia, Perth

## Controversy Over Cryptic Female Choice in Yellow Dung Flies

Leigh Simmons recounts the minor controversy he and Geoff Parker had with Paul Ward (another postdoc of Parker's). They were working on sperm utilization patterns in yellow dung flies. Parker had early on shown that there is sperm displacement in yellow dung flies. "In the new era of cryptic female choice, that was very rapidly reinterpreted it in terms of females biasing paternity toward certain males." He doesn't think there was skepticism as to whether cryptic female choice as a mechanism could operate, rather the issue was that when novel ideas come up in science, there is the "dreadful bandwagon approach", as new ideas are presented, everything is overenthusiastically reinterpreted as evidence of the new trendy phenomenon, because everybody wants to publish in *Nature* and *Science*, which is bad for science. Simmons says that is where the skepticism, or what he would rather call the healthy scientific demand for evidence, comes in, and the consideration for alternative hypotheses. He suggests that the skepticism is an inevitable consequence of the bandwagon approach. As an example, he recounts the rise and fall of the idea of fluctuating asymmetry in sexual selection—the idea that an organism's deviations from symmetry in appearance is due to environmental stress and genetic problems during development and therefore symmetry would be a measure of good genes. At first the idea was supported, but after a while there turned out to be very little support for it.

In reality, he says, what happens inside the female is a black box—cryptic female choice was an alternative explanation, but people tended to ignore mechanisms of sperm utilization. When the debates about yellow dung flies started, Simmons had been studying cryptic female choice for almost a decade, so he was convinced there were examples of cryptic female choice, but he did not think it would work in yellow dung flies, because there was sperm displacement. Entomologist Paul Ward, on the other hand, was very dedicated to the idea of cryptic female choice in yellow dung flies and suggested that females could store sperm from different males separately in their several spermatheca (sperm-storing organs). However, after many experiments, there is no evidence supporting cryptic female choice in this species.[94] Because the last male to mate displaces 80 percent of the sperm by previous mates in yellow dung flies, they are simply not a good model system to look for cryptic female choice. In species without sperm displacement, females have greater possibilities to influence which sperm fertilize their eggs.

Despite Simmons' description of no skepticism toward cryptic female choice, behavioral ecologist Charlotta Kvarnemo, who was a post-doc with Simmons in Perth, University of Western Australia when William Eberhard's *Female Control* came out in 1996, says that cryptic female choice was indeed controversial at the time.[95] In contrast to Simmons, she found that there was a quite big skepticism and resistance to cryptic female choice within the sperm competition field. Her impression was that both Leigh Simmons and Tim Birkhead, prominent figures in sperm competition research (of Simmons' publications, 175 give search hits on sperm competition and 74 on female choice[96]), were very skeptic, almost a bit nonchalant after Eberhard's book came out. She talked to Simmons about it, who expressed that although possible in principle, since cryptic female choice was almost impossible to distinguish from sperm competition empirically, it was in some way less interesting to investigate. In hindsight, she ponders, it is possible to take the other perspective—well, then it is not possible to exclude cryptic female choice if one believes oneself to study sperm competition? Possibly the discrepancy between Simmons and Kvarnemo's accounts is due a difference in attitudes toward skepticism versus healthy scientific demand for evidence, or to an omission from Simmons' part, researchers seem less willing to talk about that they have been skeptic to new ideas.

## Variability in Female Genitals

The assumption that female genitals are invariant has also contributed to the ignorance about females in sexual selection. In my interview, Leigh Simmons explains what made him reconsider the assumption that female

[94] Zuk et al., 2014: 327.

[95] Charlotta Kvarnemo interview 2017.

[96] Search on Web of Science 18 February 2022.

genitals are invariant and not subjected to sexual selection. He was working on genital morphology in dung beetles, *Onthophagus taurus*, and was interested in looking at the functional morphology (studies of how structures function) of male genitals when they found that there were female structures that the male was interacting with.[97] Here is another example of male precedence in sexual selection research.[98] The first experiment was followed up with experimental evolution studies showing divergence in those structures, and comparative analyses showed variation among species.

Simmons is critical to Eberhard's rejection, in *Sexual Selection and Animal Genitalia* (1985), of the species isolation lock-and-key hypothesis (the idea that male and female genitals evolve to a species-specific fit) based on the inference that female genitals are generally simple and without evident structures that could function as locks.[99] According to Simmons, the conclusion that female genitals are invariant was based on the taxonomic literature which is very male-biased. He explains that female genitals are difficult to investigate because in most species they consist of internal soft tissue, therefore taxonomists appropriately didn't use female genitals (which would destroy their samples) for classifying species when they could use male external genitals instead. However, recent research on female genitals has found that they too are very variable. Simmons and colleagues' studies on genital evolution in millipedes conclude that lock-and-key processes are driving evolutionary divergence in the millipede *Antichiropus variabilis*.[100] In a 2014 review on evolution of animal genitals, Simmons concludes that there is now evidence that sexual selection is the cause of widespread diversity in male genitals, but that work remains to distinguish between mechanisms of sexual selection[101] (female choice, male-male competition and sexual conflict). According to Simmons, "Sexual selection should be viewed as a continuum between male contest competition, sexual conflict and pure female choice, and it is unrealistic to expect that one particular mechanism of

[97] Ah-King, 2022

[98] Werner & Simmons, 2008.

[99] Simmons, 2014: 2.

[100] Wojcieszek & Simmons, 2012.

[101] Simmons, 2014: 14.

sexual selection will prove to be generally more important than another";[102] in addition, none of these mechanisms are mutually exclusive.[103] He says that sometimes it is difficult to distinguish between cryptic female choice and sperm competition, for example if males are varying in their sperm competitiveness and that is linked to benefits in terms of offspring survival, female multiple mating will lead to the best males fertilizing the eggs. "She doesn't need to do any cryptic female choice, just let males do their thing, facilitate the competition. So, would you call that cryptic female choice or is that…?"[104] He says there are various mechanisms by which cryptic female choice can occur—variation in whether females allow males to inseminate or not, variation in whether they store sperm or not, and if they do store sperm, whether they use sperm or not—and at each of those levels it becomes more and more difficult to demonstrate and disentangle whether it is due to male or female effects. At the early stages, observations can show whether females allow insemination or not, but at the later stages it is interpretations of a "black box".

Commenting on my colleagues' and my analysis of genital evolution,[105] he thinks we put too much emphasis on that females are understudied and on gender bias, to him the emphasis should be on interaction, it is considering the *interaction* between males and females that furthers understanding of co-evolution. I think he has a point, it is the interaction we want to understand, yet, from my perspective, the overfocus on males has led to females being understudied which is important to call attention to.

## New Technology Enables Looking into Female-Male Genital Interactions

Cryptic female choice is very hard to investigate—especially the soft tissues and female-male interactions—but the new technology of micro-computed tomography (a 3D-imaging technique that uses X-rays to make

[102] Ibid.: 14.

[103] Ibid.: 12.

[104] Leigh Simmons interview 2018.

[105] Ah-King et al., 2014.

cross-sections of an object without destroying it) enables studies of female genitals and genital interactions.[106] Because of the development of gene studies and these tomography techniques that enable looking into the "black box" of what's happening during mating, there is now some very good evidence that cryptic female choice does occur. Simmons' lab pioneered using micro-computed tomography on genital interactions in 2012.

Thus, Simmons' situated knowledges are formed by his empirical studies of sperm competition and female choice primarily in insects, his theoretical influences from Thornhill and Alcock as well as Parker, and his methodologies. He (too) problematizes the dichotomization of sperm competition and cryptic female choice and perceives the three different mechanisms in postcopulatory sexual selection—cryptic female choice, male competition and sexual conflict—as a spectrum in which it is not easy to draw the lines. Thus, he has contributed to knowledge about active females as well as males, their variable genitals and the co-evolution between males and females (Fig. 6.5).

## Postcopulatory Sexual Selection in Birds

Both sperm competition and cryptic female choice were first studied in insects, and later investigated in other animals. In my interview with ornithologist Simon Griffith, Professor at Macquarie University in Sydney, he provided his perspective on postcopulatory sexual selection research in birds. Griffith's work has developed from classical sexual selection studies on ornamentation and extra-pair paternity in European bird species to investigating pair-bonds in Australian birds and thereby challenged perceptions about females and relationships between the sexes (see Chap. 4). He has also investigated cryptic female choice in birds and describes the practical differences between studying sexual selection in insects versus birds: in insects, it is possible to observe actual copulations, measure time in copula, count spermatophores and register multiple

[106] Wojcieszek et al., 2012.

**Fig. 6.5** Photographs illustrating some study animals discussed in this chapter. **(A)** A pair of yellow dung flies, *Scathophaga stercoraria* (Bj.schoenmakers, Creative Commons Zero, Public Domain Dedication) **(B)** A scorpion fly, *Panorpa communis* (Richard Bartz, Munich aka Makro Freak, Creative Commons Attribution-Share Alike 2.5, via Wikimedia Commons) a tsetse fly *Glossina morsitans* (International Atomic Energy Agency, Creative Commons Attribution-Share Alike 4.0 via Wikimedia Commons). **(C)** A dung beetle *Onthophagus taurus* (Jayendra Chiplunkar, Creative Commons Attribution-Share Alike 4.0, via Wikimedia Commons)

mating; among birds, in contrast, much of the mating behavior is essentially invisible.[107] It is hard to observe bird copulations and what is going on in the female tracts is invisible. Griffith claims that in general researchers "don't do what they should do, they do what they can do". He thinks the problem in the sciences is that people don't say: "I want to actually find out what's going on. They start out by saying: what can we show is going on?" In birds, it is very easy to look at the male side: testes size, sperm traits, sperm performance and sperm morphology. It is a productive line of research because it is easy to do, explains Griffith, in contrast, investigating cryptic female choice is very hard. These practical

[107] Simon Griffith interview 2018.

difficulties contribute to the production of ignorance about females—that knowledge is not formed, delayed or ignored.

## Initial Neglect of Cryptic Female Choice in Bird Research

Just as in the wider field of sexual selection, male perspectives and sperm competition was the initial focus of studies of postcopulatory sexual selection in birds. In a paper depicting the history of sperm competition studies in bird research, ornithologists Tim Birkhead and Robert Montgomerie recount the starting point of sperm competition studies in birds: Birkhead was introduced to Parker's sperm competition review as an undergraduate and immediately decided to apply the idea to birds.[108] As birds were generally assumed to be monogamous, his colleagues did not believe it would succeed. However, inspired by Parker's ideas, Birkhead earliest studies of sperm competition focused on behavioral adaptations, especially frequency and timing of extra-pair copulations, and mate guarding in common guillemots *Uria aalge* (1978) and Eurasian magpies *Pica pica* (1982).[109] In 1992, Birkhead together with Danish ornithologist Anders Pape Møller[110] wrote *Sperm competition in birds*, a state-of-the-art book reviewing current knowledge and future directions. Since then, this line of research has flourished and over 1000 papers have been published on sperm competition in birds.[111]

To date, very few studies on cryptic female choice have been conducted on birds (a search on web of science shows 18 papers which concern cryptic female choice in birds[112]) and early on the idea was dismissed by ornithologists as not very important in birds. One reason for this skepticism, suggests Simon Griffith, was that comparative analyses between

[108] Birkhead & Montgomerie, 2020: 1.

[109] Ibid.: 1.

[110] Anders Pape Møller was in 2004 convicted of data fabrication by the Danish government (Vogel et al., 2004) and has traveled from one of the most prolific in sexual selection to pariah in the scientific community since he was accused of fraud (Borell, 2007).

[111] Birkhead & Montgomerie, 2020: 1.

[112] Web of science search 17 March 2022 "cryptic female choice" and bird gave 37 matches, but only 18 abstracts did concern this topic.

species of birds showed that testis size, plumage patterns and extra-pair paternity correlated. In other words, in species with elaborate plumages, there was also large testis size and high levels of extra-pair paternity. The obvious conclusion was that paternity depends on how many sperm are transferred, therefore, says Griffith, why worry about what the females are doing? There was a simple explanation.

As described in Chap. 4, early interpretations of female multiple mating and extra-pair copulation in birds were male-focused and females were perceived as passive. One example of these male-biased interpretations is Tim Birkhead's 1987 review of sperm competition in birds, in which he describes extra-pair copulations as something males impose onto females.[113] In a follow-up paper in 1990, Tim Birkhead and Fiona Hunter (also ornithologist) reviewed studies of sperm competition[114] and were criticized by William Eberhard for ignoring cryptic female choice by implicitly assuming females to be passive.[115] In their reply, Birkhead and Hunter added a description of how females may control sperm competition by behavioral, anatomical or physiological mechanisms and admitted that there is indeed a need to look at sperm competition mechanisms from a female perspective, although "this will probably not be particularly easy".[116]

Three years later, in 1993, Birkhead seems to have increased his interest in female influences on fertilization, and together with Anders Pape Møller, publish the review "Female control of paternity", pointing out behavioral and physiological mechanisms in which females might control paternity. They note that "It is becoming increasingly clear that females play a more important role than was once thought, and this is reflected in the increasing attention paid to female strategies." The same year they also publish "Why do females make it so difficult for males to fertilize their eggs?"[117]

---

[113] Birkhead, 1987: 268.

[114] Birkhead & Hunter, 1990a.

[115] Eberhard, 1990.

[116] Birkhead & Hunter, 1990b: 264.

[117] Birkhead et al., 1993.

In 2000, Tomasso Pizzari and Birkhead provided the so far most convincing evidence of cryptic female choice in birds in their study of feral fowl (feral chickens), *Gallus gallus domesticus*. They showed that females prefer mating with dominant males, and when coercively mated to subdominant males, females eject sperm and preferentially retain sperm from dominant males.[118] In a later study, Pizzari and colleagues showed that females mated with their brothers eject sperm, but retain sperm when mating with unrelated males.[119] In another of Birkhead's studies, he used artificial insemination in domestic fowl and thereby demonstrated postcopulatory cryptic female choice operating either by sperm choice or differential embryo mortality.[120] Furthermore, in 2007, Birkhead and poultry (domesticated birds) physiologist Jean-Pierre Brillard reviewed the different mechanisms of cryptic female choice in birds that could be involved in possible species barriers.[121] Poultry biologists had early knowledge about the female fertile period and that sperm from some males were disproportionally effective in fertilizing eggs[122]—another example of how knowledge about females transferred from the agricultural sciences into sexual selection research.

Also, in 2007, ornithologist Patricia Brennan with Birkhead as one of the co-authors, published a now-renowned study showing the elaborate variation in vaginal anatomy in waterfowl. They found co-evolution between vaginal anatomy and penis anatomy, and suggest that, in response to forced copulations, females have evolved a "cryptic anatomical mechanism of choice".[123]

Thus, Birkhead was first focusing on sperm competition and neglected female perspectives, then he seems to have increased his interest in cryptic female choice, outlined the possible mechanisms and experimentally explored it in several studies.

---

[118] Pizzari & Birkhead, 2000.

[119] Pizzari et al., 2004.

[120] Birkhead et al., 2004.

[121] Birkhead & Brillard, 2007.

[122] Birkhead & Montgomerie, 2020: 3–4.

[123] Brennan et al., 2007: e418.

## Ignorance of the Fertility Assurance Hypothesis

Ignorance of alternative hypotheses also contributes to the production of ignorance about females. Simon Griffith recounts that Tim Birkhead has rejected the idea of female fertility assurance (the hypothesis that females mate with extra-pair mates to assure that their eggs become fertilized) as an explanation for female multiple mating,[124] with the argument that sperm have been found on eggs that don't hatch, therefore Birkhead claimed that the male was not infertile. However, Griffith's recent research suggests that successful fertilization in zebra finches had lower numbers of sperm on the eggs than in unhatched eggs, and that interactions between the quality of the partnership and male sperm traits may influence fertilization success.[125] There seems to be an optimal number (not too few and not too many) of sperm for successful fertilization that differs between different species of birds, which suggests that even if there is sperm on eggs, there may not have been enough amounts of sperm or compatible sperm for fertilization in eggs deemed fertile by Birkhead. In Griffith's experience, it is common among researchers to dismiss the fertility assurance hypothesis on the basis of Birkhead's claims of sperm found on eggs, which contributes to this hypothesis being largely overlooked.

## Evidence of Cryptic Female Choice in Gouldian Finches

According to Simon Griffith, beyond the point of accepting or rejecting an ejaculate, there are several key points where cryptic female choice could be happening inside the female reproductive tract—which makes it difficult to demonstrate. Therefore, it has to be done in difficult roundabout ways, like counting sperm on the eggs and in sperm storage tubules. In addition, it is cumbersome to publish papers on cryptic female choice. Griffith and colleagues conducted a study on Gouldian finches, *Erythrura gouldiae*, in which there is genetic incompatibility for interbreeding between the red and black head-color morphs (if a red-headed female mates with a black-headed male

[124] See also resistance to the fertility assurance hypothesis in Chap. 4.

[125] Hurley et al., 2019.

the offspring is not viable and vice versa). They provided females—paired with compatible versus incompatible males—the opportunity for extra-pair mating with a male of the compatible or incompatible morph. A majority of females mated when the opportunity arose (irrespective of whether they were paired with a compatible or incompatible partner). However, females with incompatible social partners reduced their copulation frequency after mating with a compatible extra-pair mate. Therefore, the extra-pair matings resulted in a proportionally higher number of fertilized eggs when a female paired with an incompatible partner mated with a compatible extra-pair partner and vice versa.[126] Thus, this study showed clear evidence that cryptic female choice biased fertilizations toward genetically compatible partners. This is the second species of bird that has shown clear evidence of cryptic female choice. Yet, Griffith recounts that it would never have been published if the journal editors had not ignored a very critical review. Still, the study gets questioned whether it has proved cryptic female choice and leading researchers in the field have not cited it. Practices of not citing is yet another way in which ignorance is produced.

## Production of Ignorance About Cryptic Female Choice in Birds

Simon Griffith doesn't think that there is a heated controversy around cryptic female choice in birds, rather the idea is generally ignored. Even though researchers acknowledge that cryptic female choice occurs in birds, few try to investigate it, because it happens within the female reproductive tract, it's really hard to demonstrate beyond the stage of accepting or rejecting an ejaculate. He finds females to still be largely perceived as passive in bird research. In addition, even if the hard work of investigating cryptic female choice is done in a few studies, the number of studies looking at the male side will by far continue to outnumber them.

Thus, cryptic female choice in birds continues to be an understudied issue. Ignorance about females is produced through male precedence,

[126] Pryke et al., 2010.

greater research priority to studying sperm competition (partly because the male side is much more accessible), ignorance of alternative hypotheses, as well as the practice of not citing evidence of cryptic female choice. These mechanisms of ignorance production continue, and therefore the field of postcopulatory sexual selection in birds will continue to be male-biased. Yet, the scientific community has gone from generally dismissing cryptic female choice in birds as not important, to generally acknowledging its existence.

## Sexual Conflict: Another Example of Male Precedence

Sexual conflict occurs when evolutionary interests over reproduction differ between the sexes, for example, in the mode and frequency of mating. Although many issues in sexual selection relate to sexual conflict, the sexual conflict field (which is partly included in sexual selection, but also extends beyond it) emerged in the 1990s has focused particularly on male benefits of reproductive traits that are costly to females.[127]

Geoff Parker wrote a book chapter on sexual conflict in 1979, in which he focused on when sexual selection advantages to the male simultaneously induce costs to the female, however, he also mentioned female sexual cannibalism as another case of sexual conflict.[128] Parker's chapter received low citation rates until geneticist William Rice's noted work in the 1990s. Rice conducted experimental studies on fruit flies, *Drosophila melanogaster*, and first showed that genes could be sexually antagonistic—that is beneficial when expressed in one sex, but detrimental when expressed in the other sex.[129] He then experimentally arrested female evolution in a strain of fruit flies, while letting the male line continue to evolve, when such "arrested" females mated with "evolved" males it resulted in reduced survival of females, due to increased remating rate as

[127] Arnqvist & Rowe, 1995, Rowe & Day, 2006.

[128] Parker, 1979.

[129] Rice, 1992.

well as increased toxicity of the seminal fluid.[130] Rice's work invigorated the study of sexual conflict which has grown to be a productive field of research.

In their 2011 investigation of the field, entomologist Kristina Karlsson Green and plant ecologist Josefin Madjidian (2011) found that the reviewed theoretical models mainly investigated costs for females (as well as benefits and costs for males), and that the terminology used generally portrayed males as active and females as reactive.[131] Moreover, entomologist Melissa Plakke and colleagues assert that equivalent research on females has lagged behind, but that in all systems where females have been meticulously studied, female reproductive adaptations of sexual conflict have been identified.[132]

Furthermore, entomologist Marlene Zuk and colleagues conducted a review of the development of sexual conflict studies in 2014 showing that there was a bias in early studies toward a few insect model species which exhibit more or less intense sexual conflict, and in which sexual conflict is costly to females. As the model systems in sexual conflict diversified over time, a lower proportion of studies supported the hypothesis that sexual conflict results in harm of females.[133] Researchers tend to choose study species in which they expect to yield clear results of the investigated pattern. Species in which males are obviously harming females were chosen as the initial model systems in sexual conflict and therefore a broader diversity of study species are needed before these results can be generalizable.[134]

Although sexual conflict aims to investigate co-evolution between male and female adaptations, there has been male precedence also in this relatively new field of research.[135]

---

[130] Rice, 1996.

[131] Karlsson Green & Madjidian, 2011.

[132] Plakke et al., 2015.

[133] Zuk et al., 2014.

[134] Ibid.: 333.

[135] This section is revised and republished from Ah-King 2022 under the Creative Commons CC BY 4.0 license. https://creativecommons.org/licenses/by/4.0/

## Chapter Conclusions

Entomologist Geoff Parker's sperm competition idea (1970) was quickly accepted in the scientific community, although it received some criticism from the group selectionist's perspective. After some years of low citation rates, sperm competition became an expansive field of research. The equivalent cryptic female choice idea was suggested later (and coined in 1983 by Randy Thornhill). Cryptic female choice did not stir much interest until after the publication of William Eberhard's *Female Control: Sexual Selection by Cryptic Female Choice* (1996) in which he developed a more comprehensive framework of multiple ways in which females may influence which sperm fertilize their eggs. *Female Control* caused some controversy and resistance from researchers working on insects and birds—especially those working on sperm competition and sexual conflict (sexually antagonistic co-evolution). As evidence of cryptic female choice was established, the initial controversy gave way to the consensus that sperm competition and cryptic female choice are both valid. Lately, sperm competition and cryptic female choice have been described as a false dichotomy because all sperm competition takes place in the female reproductive tract (in internally fertilizing species). Thus, postcopulatory sexual selection is the outcome of interactions between sperm and the female environment. However, sperm competition remains much more explored than cryptic female choice, which is often more difficult to study in practice and frequently remains overlooked. Furthermore, the controversy over cryptic female choice and sexual conflict continues.

The researchers' different situated knowledges—in terms of theoretical perspectives, male embodiment ("being in a male phenotype"), influenced by their study species, engagement in male versus female perspectives—have formed their involvement in the production of knowledge and ignorance in sexual selection. Geoff Parker's understanding of sexual selection is formed by his male-focused perspective on sperm competition, influenced by sexual selection theory and his study species the yellow dung fly which exhibits fierce male-male competition and provides little opportunity for cryptic female choice. In contrast, Randy Thornhill's situated knowledges were influenced by his empirical studies of forced copulations as well as female counter-strategies in scorpion flies and hanging flies, and led him to emphasize the overlooked female perspectives in postcopulatory sexual selection. Similarly, William Eberhard's studies of the diversity of genitals among insects and spiders as well as gender-reflective approaches led him to focus on the neglected issue of cryptic female choice. Leigh Simmons's understanding of postcopulatory sexual selection was formed by his work on both sperm competition and female choice primarily in insects and new methodology to investigate the interaction between male and female genitals in action, which led him to conceptualize all three mechanisms of postcopulatory sexual selection as a spectrum.

Postcopulatory sexual selection is one example of how ignorance about females has been produced through male precedence. This chapter provides two more examples of male precedence: in genital evolution studies (in which the assumption that female genitals are invariant has contributed to ignorance about female genital evolution), whereof one example is when Leigh Simmons investigated male genital functional morphology of dung beetles and discovered that there were equivalent female structures that the male genitals interacted with, and thereafter started to investigate female genital functional morphology. In addition, the development of the relatively new field of sexual conflict provides another example of male precedence, initially focusing on male benefits and female costs, and on species in which males harm females during mating.

Other mechanisms of ignorance production include loss of acquired knowledge, for example when Geoff Parker cited a study showing that females could influence sperm viability, but missed the opportunity to develop a female perspective on postcopulatory sexual selection. The chapter also includes examples of not citing certain evidence of cryptic female choice, for example a review 20 years of cryptic female choice based the review on sperm use tests and vertrebrate studies and the fruit fly *Drosophila*, but excluded important evidence from other arthropods (insects, crustaceans, scorpions, millipedes and spiders).

When it comes to production of ignorance about females in postcopulatory sexual selection research in birds, the development followed a familiar pattern of male precedence with focusing on sperm competition first and dismissing cryptic female choice as unimportant. Although cryptic female choice is now generally acknowledged, research priorities continue to focus on sperm competition, partly because it is more accessible to study in practice. Furthermore, ignorance has been produced by unwarranted dismissal of alternative hypotheses, such as the fertility assurance hypothesis as an explanation for female multiple mating, and the practice of not citing certain evidence of cryptic female choice.

Thus, ignorance about females in postcopulatory sexual selection has been produced through the initially ignored female perspective (male precedence), the resistance toward the idea of cryptic female choice by certain researchers, practices of not citing, and the continued relatively low extent of research aimed at cryptic female choice.

## References

Ah-King, M. (2022). The history of sexual selection research provides insights as to why females are still understudied. *Nature Communications, 13*(1), 6976. https://doi.org/10.1038/s41467-022-34770-z

Ah-King, M., Barron, A., & Herberstein, M. E. (2014). Genital evolution: Why are females still understudied? *PLoS Biology, 12*(5), e1001851.

Alexander, R. D., Marshall, D. C., & Cooley, J. R. (1997). Evolutionary perspectives on insect mating. In J. C. Choe & B. J. Crespie (Eds.), *The evolution of mating systems in insects and arachnids*. Cambridge University Press.

Arnqvist, G., & Rowe, L. (1995). Sexual conflict and arms races between the sexes: A morphological adaptation for control of mating in a female insect. *Proceedings of the Royal Society B: Biological Sciences, 261*, 123–127.

Birkhead, T. R. (1987). Sperm competition in birds. *Trends in Ecology & Evolution, 2*, 268–272.

Birkhead, T. R. (1996). In it for the eggs. *Nature, 382*, 772.

Birkhead, T. R. (1998a). Sperm competition in birds: Mechanisms and function. In T. R. Birkhead & A. P. Møller (Eds.), *Sperm competition and sexual selection* (pp. 579–622). Academic Press.

Birkhead, T. R. (1998b). Cryptic female choice: Criteria for establishing female sperm choice. *Evolution, 52*(4), 1212–1218.

Birkhead, T. R. (2000a). *Promiscuity: An evolutionary history of sperm competition and sexual conflict*. Faber.

Birkhead, T. R. (2000b). Defining and demonstrating postcopulatory female choice—Again. *Evolution, 54*(3), 1057–1060.

Birkhead, T. R., & Brillard, J.-P. (2007). Reproductive isolation in birds: Postcopulatory prezygotic barriers. *Trends in Ecology & Evolution, 22*(5), 267–272.

Birkhead, T. R., Chaline, N., Biggins, J. D., Burke, T., & Pizzari, T. (2004). Nontransitivity of paternity in a bird. *Evolution, 58*, 416–420.

Birkhead, T. R., & Hunter, F. M. (1990a). Mechanisms of sperm competition. *Trends in Ecology & Evolution, 5*(2), 48–52.

Birkhead, T. R., & Hunter, F. M. (1990b). Reply to from Tim Birkhead and Fiona Hunter. *Trends in Ecology & Evolution, 5*(8), 263–264.

Birkhead, T. R., & Møller, A. P. (1992). *Sperm competition in birds: Evolutionary causes and consequences*. Academic Press.

Birkhead, T. R., & Møller, A. P. (1998). *Sexual selection and sperm competition*. California Academic Press.

Birkhead, T. R., Møller, A. P., & Sutherland, W. J. (1993). Why do females make it so difficult for males to fertilize their eggs? *Journal of Theoretical Biology, 161*, 51–60.

Birkhead, T. R., & Montgomerie, R. (2020). Three decades of sperm competition in birds. *Philosophical Transactions of the Royal Society B, 375*, 20200208. https://doi.org/10.1098/rstb.2020.0208

Borell, B. (2007). A fluctuating reality. *The Scientist*, January 1.

Brennan, P. L. R., Prum, R. O., McCracken, K. G., Sorenson, M. D., Wilson, R. E., & Birkhead, T. R. (2007). Coevolution of male and female genital morphology in waterfowl. *PLoS One, 2*(5), e418. https://doi.org/10.1371/journal.pone.0000418

Eberhard, W. G. (1985). *Sexual selection and animal genitalia*. Harvard University Press.

Eberhard, W. G. (1990). Inadvertent machismo? *Trends in Ecology & Evolution, 5*(8), 263.

Eberhard, W. G. (1996). *Female control: Sexual selection by cryptic female choice*. Princeton University Press.

Eberhard, W. G. (1998). Female roles in sperm competition. In T. R. Birkhead & A. P. Møller (Eds.), *Sperm competition and sexual selection* (pp. 91–116). Academic Press.

Eberhard, W. G. (2000). Criteria for demonstrating postcopulatory female choice. *Evolution, 54*(3), 1047–1050.

Eberhard, W. G. (2015). Cryptic female choice and other types of post-copulatory sexual selection. In A. V. Peretti & A. Aisenberg (Eds.), *Cryptic female choice in arthropods: Patterns, mechanisms and prospects*. Springer.

Eberhard, W. G. (2017). Cryptic female choice revisited: A response to Firman et al. *Trends in Ecology & Evolution, 32*(11), 805–806.

Eberhard, W. G., & Lehmann, G. U. C. (2019). Demonstrating sexual selection by cryptic female choice on male genitalia: What is enough? *Evolution, 73–12*, 2415–2435.

Firman, R. C., Gasparini, C., Manier, M. K., & Pizzari, T. (2017a). Postmating female control: 20 years of cryptic female choice. *Trends in Ecology & Evolution, 32*, 368–382.

Firman, R. C., Gasparini, C., Manier, M. K., & Pizzari, T. (2017b). Cryptic female choice: A general phenomenon. A reply to Eberhard. *Trends in Ecology & Evolution, 32*(11), 807.

Gowaty, P. A. (1982). Sexual terms in sociobiology—Emotionally evocative and paradoxically, jargon. *Animal Behaviour, 30*, 630–631.

Haraway, D. J. (1988). Situated knowledges: The science question in feminism and the privilege of partial perspectives. *Feminist Studies, 14*(3), 575–599.

Haraway, D. J. (1989). *Primate visions: Gender, race and nature in the world of modern science*. Routledge.

Holland, B., & Rice, W. R. (1997). Cryptic sexual selection—More control issues. Review of William Eberhard's Female control: Sexual selection by cryptic female choice. *Evolution, 51*(1), 321–324.

Hurley, L. L., Rowe, M., & Griffith, S. C. (2019). Reproductive coordination breeds success: The importance of the partnership in avian sperm biology. *Behavioral Ecology and Sociobiology, 74*(1), 3. https://doi.org/10.1007/s00265-019-2782-9

Karlsson Green, K., & Madjidian, J. A. (2011). Active males, reactive females: Stereotypic sex roles in sexual conflict research? *Animal Behaviour, 81*, 901–907. https://doi.org/10.1016/j.anbehav.2011.01.033

Kirkpatrick, M., & Ryan, M. J. (1991). The evolution of mating preference and the paradox of the lek. *Nature, 350*, 33–38.

Lloyd, J. E. (1979). Mating behavior and natural selection. *Florida Entomologist, 62*, 17–34.

Lüpold, S., Reil, B., Jonathan, M., Mollie, K., Zeender, V., Belote, J. M., & Pitnick, S. (2020). How female × male and male × male interactions influence competitive fertilization in *Drosophila melanogaster*. *Evolution Letters, 4–5*, 416–429.

Malin, Ah-King (2022). The history of sexual selection research provides insights as to why females are still understudied. *Nature Communications 13*(1) 6976 10.1038/s41467-022-34770-z

McDonough-Goldstein, C. E., Pitnick, S., & Dorus, S. (2022). *Drosophila* female reproductive glands contribute to mating plug composition and the timing of sperm ejection. *Proceedings of the Royal Society of London B, 289*, 20212213. https://doi.org/10.1098/rspb.2021.2213

Moore, J. C., & Pannell, J. R. (2011). Sexual selection in plants. *Current Biology, 21*(5), PR176–PR182. https://doi.org/10.1016/j.cub.2010.12.035

Olzer, R., Ehrlich, R. L., Heinen-Kay, J. L., Tanner, J., & Zuk, M. (2018). Reproductive behavior. In A. Córdoba-Aguilar, D. González-Tokman, & I. González-Santoyo (Eds.), *Insect behavior: From mechanisms to ecological and evolutionary consequences* (pp. 189–202). Oxford University Press.

Parker, G. A. (1970a). The reproductive behaviour and the nature of sexual selection in Scatophaga stercoraria L. (Diptera: Scatophagidae): V. The female's behaviour at the oviposition site. *Behaviour, 37*(1–2), 140–168.

Parker, G. A. (1970b). Sperm competition and its evolutionary consequences in the insects. *Biological Reviews, 45*, 525–567.

Parker, G. A. (1970c). The reproductive behaviour and the nature of sexual selection in Scatophaga stercoraria L. (Diptera: Scatophagidae): I. Diurnal and seasonal changes in population density around the site of mating and oviposition. *Journal of Animal Ecology, 39*(1), 185–204.

Parker, G. A. (1979). Sexual selection and sexual conflict. In M. S. Blum & N. A. Blum (Eds.), *Sexual selection and reproductive competition in insects* (pp. 123–166). Proceedings International Congress of Entomology.

Parker, G. A. (1998). Sperm competition and the evolution of ejaculates: Towards a theory base. In T. R. Birkhead & A. P. Møller (Eds.), *Sperm competition and sexual selection* (pp. 3–53). Academic Press.

Parker, G. A. (2001). Golden flies, sunlit meadows: A tribute to the yellow dungfly. In L. A. Dugatkin (Ed.), *Model systems in behavioural ecology: Integrating conceptual, theoretical, and empirical approaches* (pp. 3–26). Princeton University Press.

Parker, G. A. (2010a). In celebration of questions, past, present and future. In T. Székely, A. J. Moore, & J. Komdeur (Eds.), *Social behaviour: Genes, ecology and evolution* (pp. 267–270). Cambridge University Press.

Parker, G. A. (2010b). Reflections before dusk. In L. C. Drickamer & D. A. Dewsbury (Eds.), *Leaders in animal behavior, second generation* (pp. 429–464). Cambridge University Press.

Peretti, A. V., & Aisenberg, A. (Eds.). (2015). *Cryptic female choice in arthropods: Patterns, mechanisms and prospects*. Springer International Publishing.

Pitnick, S., & Brown, W. D. (2000). Criteria for demonstrating female sperm choice. *Evolution, 54*(3), 1052–1056.

Pizzari, T., & Birkhead, T. R. (2000). Female feral fowl eject sperm of subdominant males. *Nature, 405*(6788), 787–789.

Pizzari, T., Løvlie, H., & Cornwallis, C. K. (2004). Sex-specific, counteracting responses to inbreeding in a bird. *Proceedings of the Royal Society B: Biological Sciences, 271*, 2115–2121.

Plakke, M. S., Deutsch, A., Meslin, C., Clark, N. L., & Morehouse, N. I. (2015). Dynamic digestive physiology of a female reproductive organ in a polyandrous butterfly. *Journal of Experimental Biology, 218*(10), 1548–1555.

Pryke, S. R., Rollins, L. A., & Griffith, S. C. (2010). Females use multiple mating and genetically loaded sperm competition to target compatible genes. *Science, 329*(5994), 964–967.

Rice, W. R. (1992). Sexually antagonistic genes: Experimental evidence. *Science, 256*, 1436–1439.

Rice, W. R. (1996). Sexually antagonistic male adaptation triggered by experimental arrest of female evolution. *Nature, 361*, 232–234.

Richardson, S. S. (2013). *Sex itself: The search for male and female in the human genome*. University of Chicago press.

Rose, H., & Rose, S. P. R. (Eds.). (2000). *Alas, poor Darwin: Arguments against evolutionary psychology*. Harmony Books.

Rowe, L., & Day, T. (2006). Detecting sexual conflict and sexually antagonistic coevolution. *Philosophical Transactions of the Royal Society of London B. Biological Sciences, 361*, 277–285.

Schilthuizen, M. (2014). *Nature's Nether Regions: What the sex lives of bugs, birds, and beasts tell us about evolution, biodiversity, and ourselves*. Penguin Books.

Seeley, T. D., & Sherman, P. W. (2021). Adaptive design. Retrieved February 9, 2021, from https://www.britannica.com/science/animal-behavior/Adaptive-design#ref1043231

Shuker, D. M., & Simmons, L. W. (Eds.). (2014). *Evolution of insect mating systems. Preface*. Oxford University Press.

Simmons, L. W. (1986). Female choice in the field cricket *Gryllus bimaculatus* (De Geer). *Animal Behaviour, 34*(5), 1463–1470.

Simmons, L. W. (2001). *Sperm competition and its evolutionary consequences in the insects* (Monographs in behavior and ecology). Princeton University Press.

Simmons, L. W. (2014). Sexual selection and genital evolution. *Australian Entomology, 53*, 1–17.

Small, M. F. (Ed.). (1984). *Female primates: Studies by women primatologists*. Alan R. Liss.

Smith, J. M. (1991). Theories of sexual selection. *Trends in Ecology & Evolution, 6*, 146–151.

Smith, R. L. (Ed.). (1984). *Sperm competition and the evolution of animal mating systems*. Academic Press.

Tallamy, D. W. (1997). Review: Female control: Sexual selection by cryptic female choice by William G. Eberhard. *American Scientist, 85*(5), 485–486.

Thornhill, R. (1976). Sexual selection and nuptial feeding behavior in *Bittacus apicalis* (Insecta: Mecoptera). *American Naturalist, 110*(974), 529–548.

Thornhill, R. (1980). Rape in *Panorpa* and a general rape hypothesis. *Animal Behaviour, 28*, 52–59.

Thornhill, R. (1983). Cryptic female choice and its implications in the Scorpionfly *Harpobittacus nigriceps*. *American Naturalist, 122*(6), 765–788.

Thornhill, R. (1984). Alternative hypotheses for traits believed to have evolved by sperm competition. In R. L. Smith (Ed.), *Sperm competition and the evolution of animal mating systems*. Academic Press.

Thornhill, R. (2015). Cryptic female choice: A tale about a boy who loved flies. In A. V. Peretti & A. Aisenberg (Eds.), *Cryptic female choice in arthropods: Patterns, mechanisms and prospects*. Springer International Publishing.

Thornhill, R., & Alcock, J. (1983). *The evolution of insect mating systems* (p. 1983). Harvard University Press.

Thornhill, R., & Palmer, C. (2000). *A natural history of rape: Biological bases of sexual coersion*. MIT Press.

Travis, C. B. (Ed.). (2003). *Evolution, gender, and rape: Multidisciplinary critiques of the notion of rape as an evolutionary adaptation*. MIT Press.

Vogel, G., Proffitt, F., & Stone, R. (2004). Ecologists rocked by misconduct finding. *Science, 28*. https://doi.org/10.1126/article.34801

Waage, J. K. (1979). Dual function of the damselfly penis: Sperm removal and transfer. *Science, 203*(4383), 916–918. https://doi.org/10.1126/science.203.4383.916

Werner, M., & Simmons, L. W. (2008). The evolution of male genitalia: Functional integration of genital sclerites in the dung beetle *Onthophagus taurus*. *Biological Journal of the Linnean Society, 93*, 257–266.

Wojcieszek, J. M., Austin, P., Harvey, M. S., & Simmons, L. W. (2012). Micro-CT scanning provides insight into the functional morphology of millipede genitalia. *Journal of Zoology, 287*(2), 91–95. https://doi.org/10.1111/j.1469-7998.2011.00892.x

Wojcieszek, J. M., & Simmons, L. W. (2012). Evidence for stabilizing selection and slow divergent evolution of male genitalia in a millipede (*Antichiropus variablilis*). *Evolution, 66*(4), 1138–1153. https://doi.org/10.1111/j.1558-5646.2011.01509.x

Zuk, M., Garcia-Gonzalez, F., Herberstein, M. E., & Simmons, L. W. (2014). Model systems, taxonomic bias, and sexual selection: Beyond *Drosophila*. *Annual Review of Entomology, 59*, 321–338.

# 7

# Conclusions: An Epistemology of Ignorance in Sexual Selection

During the last decades of the twentieth century, evolutionary biologists began shifting their understanding of females from being "coy", passive and monogamous toward acknowledging that females can have active sexual strategies, be competitive, frequently mate with multiple males and are variable among themselves—a shift I call a *female turn*. The history of this female turn shows the heterogeneous histories of and multiple aspects of shifting perceptions about females in the different subfields concerning research on primates, birds, insects, snakes, frogs, fish and spiders. My analysis combines the theoretical frameworks of situated knowledges and epistemology of ignorance to show who knew and who did not know about active and multiply mating females, and why. The researchers' theoretical perspectives, feminist insights, personal experiences, questions and specific study species formed their particular situated knowledges and ignorance about sexual selection, as well as their involvement in controversies and negotiations of scientific knowledge. Thereby, this history demonstrates how ignorance about females has been produced, maintained and resisted, and ways in which females continue to be neglected in the field.

In this concluding chapter, I first discuss the use of the turn metaphor, then summarize my analysis of the researchers' situated knowledges and



M. Ah-King, *The Female Turn*, https://doi.org/10.1007/978-981-19-7161-7_7

ignorance production about females in sexual selection research throughout the history of the female turn. I describe the different mechanisms of ignorance production that I have identified. Furthermore, I present an alternative definition of a female turn in sexual selection, namely delineated by the scientific community's knowledge about sexual selection *in females*. According to this characterization, the female turn has merely begun, because knowledge about sexual selection in females is still rudimentary. I examine what presently hinders research on sexual selection in females and reflect on the relations of gender and science. Finally, I scrutinize my own situated knowledges—how my position as an evolutionary biologist and gender researcher, my theoretical perspectives and experiences have formed my understanding of the female turn.

## The Turn Metaphor

All history writing concerns the questions of continuity and change. So, when does one stipulate that a turn has occurred? Historical studies have often declared turns, such as a "linguistic turn" in philosophy.[1] A turn may indicate a trend, for example, the tendency to focus on the role of animals in literary studies[2] has been depicted as an "animal turn", even though not everyone in the field participated in it. Historian Dan Vandersommers distinguishes between a turn and a subfield, and designated an "animal turn" in historical studies, "because animals have slithered into history from many directions, animal history represents a 'turn', not a specific subfield".[3] However, the use of the turn terminology has been criticized. Anthropologist Gary Wilder argues that the turn metaphor makes tendencies seem more homogeneous than they were in reality.[4] Similarly, feminist theorist Clare Hemmings has criticized the use of the wave metaphor in feminist historiography, because it is used to foreground a story about progress that emphasizes change and downplay continuity, thereby making invisible historical multiplicities.[5]

[1] Glock & Kalhat, 2018.

[2] See for example Lönngren, 2015.

[3] Vandersommers, 2016.

[4] Wilder, 2012.

[5] Hemmings, 2011.

Since perceptions about females have shifted in many different subfields of evolutionary biology and animal behavior, and the new studies that include active, multiply mating and variable females are conducted within these subfields, I consider it motivated to describe this occurrence as a (multifaceted) female turn. By emphasizing the various situated knowledges and the ways in which ignorance is produced in my analysis of the *female turn*, I highlight multiplicities and variability—thus continuity as well as change in this historiography of sexual selection.

## Situated Knowledges and Ignorance Production in Sexual Selection

In a general sense, all knowledge is situated, because it is not possible to perform what science theorist Donna Haraway calls the "god trick of seeing everything from nowhere".[6] Researchers are always part of their cultural and social context which influences the production of science. Because of the awareness of biases, natural science methodology aims at scientific objectivity and to avoid biases by describing methods in detail, so that the observer can be replaced by any other observer. Yet, science historians have revealed social and cultural biases in science, and that science communities' exclusion of people based on gender, class, sexuality and race have influenced the knowledge produced[7] and not produced.[8]

This analysis of the female turn is an epistemology of ignorance about females in sexual selection. It shows who knew and who did not know about female sexual agency and why I investigate what prevented scientists from engaging in the question of female sexual agency and what made certain researchers challenge assumptions about coy and passive females and forward new perceptions about females and their agency. The analysis demonstrates how the researchers' "specific *ways* of seeing"[9] or "prosthetic technologies interfaced with our biological eyes

[6] Haraway, 1988: 581.

[7] Harding, 1986, Haraway, 1988, Schiebinger, 1989.

[8] Tuana, 2006, Proctor & Schiebinger, 2008.

[9] Haraway, 1988: 583.

and brains"[10] form their situated knowledges and ignorance[11]—through their human physiological senses, the specific animals they are studying, the theoretical perspectives they employ, their methodologies, technologies, their institutional position, their geographical location, their feminist insights or engagement in gender-reflective approaches, as well as their sex/gender and lived experiences. These specific circumstances have shaped their different understandings of sex and sexual selection.

Importantly, scientific knowledge is not an individual undertaking, but as science philosopher Helen Longino asserts, knowledge is constructed by "an interactive dialogic community,"[12] or in science philosopher Lynn Hankinson Nelson's words, "what we know is inextricably dependent on knowledge, standards, and practices of the epistemic communities that we are members of."[13] This history of the female turn shows that perceptions about females have shifted at different times in the different subfields of evolutionary biology—as a result of the researchers' specific situated knowledges and ignorance, and negotiations of scientific knowledge, even though the subfields are all part of the larger scientific community of evolutionary biology.

Though both male and female animals have always been present in the sexual selection studies and in the natural history observations that led Charles Darwin to develop his theory, this history of sexual selection shows historical and ongoing ignorance production about females—knowledge about females has been ignored, delayed and not formed.

## Formulation of the Theory

Charles Darwin's situated knowledges and ignorance about sexual selection was formed by his observations and experiences in his specific time

[10] Ibid.: 589.

[11] Tuana, 2006: 3.

[12] Longino, 2006: 272.

[13] Nelson, 1993: 52.

and society. He formulated his theory both within and against prevailing Victorian ideas about of sex, gender and race.[14] His theorizing was influenced by his understanding of the art of bird breeding, his experiences traveling around the globe as a gentleman naturalist, the multitude of information gained by his many correspondents, his theories of embryogenesis (that the male was the more developed sex) and inheritance (that traits that appear in either sex late in life develop in the same sex alone, while traits that appear early in life usually develop in both sexes[15]), as well as Victorian societal ideals. Furthermore, his personal experiences did not challenge his Victorian ideas about empathic, mothering and caring women and genial (white, mostly upper-class) men.[16] Despite this background, he contradicted Victorian ideas about females, when he cast female animals (but not women) in the active role of choosing mates, although in a rather passive and reluctant way. Thus, even though he was convinced of female choice, Victorian ideals pervaded Darwin's generalizations of females as coy and passive in sexual encounters.

Furthermore, human visual and auditory biases combined with the common biological pattern of male extravagant visual and auditory characteristics (especially in birds, and prevalent male birdsong in the Northern hemisphere) and Darwin's difficulty in explaining them by his theory of natural selection (selection for survival benefits) resulted in sexual selection theory's primary focus on the evolution of male extraordinary traits. This androcentrism[17] influenced his formulation of sexual selection theory, and has continued to affect and produce ignorance about females in subsequent research in the field.

## Reception of the Theory

Scientists contemporary to Darwin, although acknowledging male-male competition (but subsuming it under natural selection), did not accept female choice as a mechanism for sexual selection in animals. In line with prevailing Victorian ideology, they perceived females as passive, and

[14] Richards, 2017.

[15] Richards, 2017: 284.

[16] Ibid.: 36.

[17] Androcentrism is the practice of placing the male point of view in the center of one's worldview.

furthermore, they did not think animals had the aesthetic sensibility or rational ability they presumed was needed in order to choose among different mates.[18] Darwin's co-discoverer of natural selection, Alfred Russel Wallace, formed his particular view of sexual selection through his experience of tropical expeditions collecting insects and butterflies and his studies of evolution of camouflage through natural selection. To Wallace, females had evolved their dull plumages through natural selection for protection against predation, and there was no need to explain male extraordinary plumages through female choice. Therefore, he largely dismissed sexual selection, as did most in the scientific community, except for the potential of female choice in humans.

A number of (American, white, educated and middle-class) women were engaged in evolutionary theory in the last decades of the nineteenth century. Because male superiority in society was presumed to be instituted by God, they turned to science to claim gender equality. They criticized Darwin for describing women as intellectually inferior and also that he ignored the evolution of female characteristics.[19] They argued that male choice in humans was unnatural, since, according to Darwin, female choice of mates was the norm among animals. However, the subsequent professionalization and masculinization of science excluded them and their feminist critique. In the following decades, biologists in the ensuing all-male scientific community showed sparing interest in sexual selection—except for its application to humans (where informed mate choice was by eugenicists and a popular audience seen as a method of improving humanity).[20]

[18] Ibid.: 523ff.

[19] Hamlin, 2014.

[20] Milam, 2010: 43ff.

## Forming the Sexual Selection Field

When sexual selection emerged as a field of study in the 1970s, stereotypical perceptions of females as passive and coy were sustained. Human visual and auditory biases influenced which animals were studied in sexual selection. Thus, birds and insects came to dominate sexual selection studies. Investigations of male extravagant characteristics, such as those in the long-tailed widow birds, preceded those of female ornamentation, in for example pipefish. This is one example of male precedence—that research starts with male-centered investigations or explanations and later includes the female-centered equivalents.[21]

## Early Female Turn in Primatology

Primates play a special role in human imagination as a mirror and model for human evolution. In the Japanese context, this meant that primate studies focused on trying to figure out how the whole "species society" and social relations functioned. Japanese primatologists already discovered in the 1950s that dominance hierarchies in Japanese macaques was mostly dependent on strong kinship bonds along female relatives.[22] Thus, there does not seem to have been a female turn in Japanese primatology, because they did not assume females to be passive in the first place.

In contrast, early western male-dominated primate studies focused mainly on male dominance and aggressive interactions while females were considered as resources to males and in their role as mothers. Although there were a few studies that described females as active, the prevailing theoretical model of primate society focused on male domination and aggression. Pioneering feminist primatologists noted the discrepancy between the theoretical assumptions of coy and passive females and their observations of sexually initiating and actively strategizing females. Sarah Blaffer Hrdy describes how her empathy for langur females that got their infants killed by males

[21] Ah-King, 2022.
[22] Haraway, 1989: 253.

engendered new research questions that put females at center stage. Thus, she paid special attention to female behavior and found that females have counter-strategies to infanticide by males. Her position at Harvard also placed her in the middle of the emerging field of sociobiology—as well as the shift from "good-for-the-species" to individual benefits explanations of behavior—which formed her interpretation of infanticide by males as a sexually selected strategy. Her observations and gender-reflective perspectives led her to question assumptions about coy females, self-sacrificing mothers as well as the hypothesis that the female orgasm evolved as a by-product of the evolution of male orgasm. She was not alone, in the time after the sexual revolution and successes of the feminist movement, a number of pioneering feminist primatologists' interventions lead to methodological refinement, investigations of females' lives and forwarded new perceptions of females as active, aggressive, dominant, and pursuing their own strategies.[23] These interventions have impacted the field of primatology way beyond feminist circles, to question prevailing theory, shift methods, pay attention to females' lives and strategies, aggression and dominance, and develop models framing females as active has become mainstream.[24] Furthermore, although western primatologists initially disregarded the findings of Japanese primatology, and thereby delayed the transmission of knowledge, these findings of active and powerful females have also been incorporated.

## Bird Research

Birds have been described as the "paradigmatic" study species in the field—they exhibit the extraordinary male ornaments that puzzled Darwin in the first place. As the research focus was on understanding the evolution of such extravagant male traits, Swedish ornithologist Malte Andersson—inspired by current theoretical and empirical influences—pioneered field studies of female choice with a cut-and-glue-experiment on long-tailed widow birds (*Euplectes progne*). Andersson's particular knowledges about sexual selection were formed by his work with specific bird species, his mathematical modeling and current theoretical debates.

[23] Haraway, 1989.

[24] Fedigan, 2001.

Thus, his experiment contributed to the ending of the controversy over female choice as a mechanism of sexual selection. Yet, perceptions about females as coy and passive lingered. As 90 percent of bird species are socially monogamous—and in line with Angus Bateman's prediction that females don't benefit by mating with multiple males—ornithologists assumed females to mate with one male.

With the development of new molecular techniques that could determine who was producing which offspring, there was a surge of studies revealing frequent "extra-pair paternity", that is female multiple mating, even in supposedly monogamous birds. In line with prevalent perceptions of coy and passive females, initial interpretations of this unexpected multiple mating in females were male-focused—males were thought to harass or force copulations onto females in order to gain more offspring.[25] The large number of behavioral observations of species with forced extra-pair copulations[26] was one basis for interpreting extra-pair copulations as caused by males. However, there exist very early (1937) natural history accounts of females initiating or readily accepting to mate with other males than their social mate,[27] but these interpretations were ignored or re-interpreted as (possibly) "apparent willingness" rather than "real."[28] Thus, accounts of active females did not affect the consensus interpretation of female multiple mating until the mid-1990s.[29] Thereafter, due to feminist ornithologists' challenges and studies, there has been a shift toward interpreting female multiple mating as an active female strategy, or at least acknowledging that females can have active sexual strategies. Thus, the interpretations of female multiple mating provide an example of production of ignorance through male precedence.

The study species is an important influence on the researchers' understanding of sex and sexual selection. For example, guillemot females (*Uria aalge*) with short tails, living in high-density colonies cannot hinder

[25] McKinney et al., 1984; Birkhead, 1987.

[26] McKinney et al., 1984.

[27] Hann, 1937.

[28] McKinney et al., 1984.

[29] Burley & Parker, 1998: 2.

forced copulations, whereas razorbill females (*Alca torda*), with long tails, can prevent unwanted copulations.[30] Studying these species has formed the researchers' understandings—Tim Birkhead and coworkers, who studied guillemots, described females as victims of extra-pair copulations,[31] while Richard Wagner studying razorbills described females as active.[32] However, a later study of guillemots showed female agency: in (the fewer) instances when females were initiators, extra-pair copulations often led to cloacal contact and therefore more likely produced offspring.[33] Thus, although the species is an important influence, other factors also form the researchers' "ways of seeing" sexual selection.

Motivated by feminist insights, ornithologist Patricia Gowaty questioned assumptions about passive females and devised experiments to investigate birds with the assumption that females are active. Her institutional position provided collegial interest and skills in parentage studies to reveal multiple paternity and maternity in socially monogamous eastern bluebirds (*Sialia sialis*) and she pioneered studies of female-female aggression in birds. She has also criticized the prevailing Darwin-Bateman paradigm in sexual selection, repeated Bateman's classic experiment and developed gender-neutral models of mate choice. Susan Smith also presented feminist critique questioning prevailing assumptions that females were passive in extra-pair copulations. Radio-tracking (a new technology that enabled following individual birds in the field) female black-capped chickadees (*Poecile atricapillus*), she found that they actively sought out high-ranked males for extra-pair copulations.[34] Thus, although feminists' cultural maps of females—as actively soliciting matings, influencing which sperm fertilize their eggs, being variable among themselves, sometimes aggressive and dominant—still meet resistance and are not fully acknowledged in the scientific community, general perceptions of female birds now incorporate possibilities of active female strategies.

---

[30] Wagner, 1991: 163.

[31] Birkhead et al., 1985.

[32] Wagner, 1991.

[33] Hatchwell, 1988.

[34] Smith, 1988.

Simon Griffith's studies on male ornaments in classical European bird model species and his shift to working on long-lived Australian birds made him question prevalent views that sexual selection mainly concerns selection for extraordinary male ornaments. Instead, he emphasizes the importance of the quality of the pair-bond in many Australian bird species—the cooperation between the pair is crucial for their success in producing viable offspring—another example of the influence of geographical location and its species. Furthermore, he alleges that the fertility assurance hypothesis is overlooked, that is the hypothesis that low levels of extra-pair paternity may be due to females' active strategy to assure fertilization of their eggs. This argument is part of a continuing controversy over interpretations of female birds as active or passive in extra-pair mating and sexual conflict experiments.

Thus, perceptions about female birds in the scientific community of bird researchers have shifted from monogamous to acknowledging frequent multiple mating, and from passive to including possibilities of female active strategies.

## Insect Research

In the 1950s, agricultural pest-control studies using the sterile male technique revealed that females mated with multiple males in many insect species. Likewise did natural history studies and population genetic studies (which investigated the maintenance of genetic variation by mate choice in the 1960s). Thus, among insect researchers, female insects were acknowledged to mate with multiple males at the revival of sexual selection research in the 1970s. In fact, one might even say that the revival started on the premise of female multiple mating—it was the point of departure for entomologist[35] and theoretical biologist Geoff Parker's formulation of the sperm competition hypothesis (1970).[36] Yet, the questions he was interested in and the experiments he found feasible mainly concerned males. Parker's situated knowledges, formed by his use of yel-

[35] Entomologists are scientists studying insects.

[36] Parker, 1970.

low dung flies as models—a system dominated by male competition with little room for female influence over which males fertilize their eggs—his primarily male perspective and theoretical models, have been and still are very influential in contemporary sexual selection.

Some years later, entomologist Randy Thornhill and his colleagues suggested the equivalent female-centered hypothesis in post-copulatory sexual selection,[37] cryptic female choice (females influence on which sperm fertilize their eggs after mating with several males). Thornhill came to pay attention to the female strategies counteracting forced copulations in his study species the scorpionfly *Harpobittacus nigriceps*. He noted that females determined the length of copulation in relation to the size of the nuptial gift they were offered by males—a discovery that lead him to coin the term cryptic female choice (1983).[38] Furthermore, tropical biologist William Eberhard has broadened the understanding of cryptic female choice as a general phenomenon through challenging prevailing assumptions and compiling findings from several disparate fields—morphology, reproductive physiology and behavioral studies. Eberhard's situated knowledges are formed by his empirical studies of insects and spiders, his gender-reflective perspectives on male bias in previous interpretations and theoretical perspectives including a multitude of ways in which females may exercise cryptic choice. Thus, both Thornhill and Eberhard have advocated for perceptions of females as active.

Furthermore, entomologist Darryl Gwynne pioneered the research on how the sexes can flexibly shift between competing for mates and mate selectively—depending on their relative parental investment, population density and food availability. The katydid insects provided a model for perceiving the sexes as flexible and sexual selection as dynamic, which opened up for perceptions of females as competitive and active, at least in some unusual species.

In addition, entomologist Leigh Simmons started working with female choice in crickets in which males provide nutrition to the female at mating, and then broadened his research to include sperm competition and

[37] The development of these ideas was "kind of collective" with James Lloyd and John Sivinski, see Chap. 6.

[38] Thornhill, 1983.

genital evolution. His particular knowledges were formed by his empirical studies of sperm competition and female choice primarily in insects, current theoretical influences from both Thornhill and Parker, and his specific methodologies. He problematizes the dichotomization of sperm competition and cryptic female choice and perceives the three different mechanisms in postcopulatory sexual selection—cryptic female choice, male competition and sexual conflict—as a spectrum in which it is not easy to draw the lines. Thus, he has contributed to knowledge about active females as well as males, their variable genitals and the co-evolution between males and females.

In the community of insect researchers, female multiple mating was widely recognized early on, but perceptions of females as passive dominated in studies of insect sexual selection. These perceptions have shifted toward acknowledging female agency through the development of cryptic female choice studies (although still underrepresented), studies of species in which males provide females with nutrition at mating, and species that are unusually sexually dynamic in their level of parental investment and competition.

## Snake Research

Also in snake research, the study species have influenced the researchers' situated knowledges. The secretive nature of snakes—they are hard (for a human) to follow in the field, they hide underground and in trees, they hibernate in cold habitats—means that their mating patterns are difficult to investigate. In addition, because snakes primarily use chemical communication rather than visual and auditory signals, sexual selection in snakes is not obvious to humans. Male snakes are also very choosy, contrary to general assumptions of sexual selection. Thus, according to herpetologist Richard Shine, natural history observations and investigations of sex differences in size among snakes were not guided by prevalent assumptions about strong sexual selection on gaudy and displaying males—but instead investigated the respective influences of natural and sexual selection on both males and females.

Shine's situated knowledges about the sexes and sexual selection are shaped by his natural history observations of snakes in the field, his unusually broad taxonomic overview as well as gender-reflective perspectives. Thus, Shine has contributed to the shift in perceptions about females, away from notions of females as passively mating with courting snake males to actively benefiting from their multiple mating strategies, showing female toads' capabilities for actively deterring courting males, but also revealing that the seemingly sensual courtship of male snakes can physiologically function to force females to open their cloaca for mating. Furthermore, he has highlighted the common ecological causes of sex differences in contrast to predominant sexual selection explanations.

Because sexual selection in snakes is not as obvious to a human observer, the research on snakes has not suffered the same male bias as mammal and bird studies. Yet, Jesús Rivas and his colleague's research lead to a shift in perceptions of female multiple mating also in snake research (2005). Again the study species was important in forming Rivas' particular view, as well as his critical perspectives on how different biases influence research in animal behavior—the bias of focusing on snakes in temperate zones thus overlooking tropical biology, and the bias of "unthinking anthropomorphism" constraining our understanding of other animals. His field studies on green anacondas lead to his discovery of female multiple mating and later to deduce, from a reinterpretation of earlier studies, that female multiple mating is widespread among snakes and had been neglected in categorizations of snake mating systems as generally polygynous.

## Frog Research

Because frogs have external fertilization, perceptions of female multiple mating have not shifted much—there was no controversy around females mating with several males. Yet, paternity analyses have revealed more "extra-pair" paternity and more intense sperm competition than formerly presumed.

Mike Ryan's research on female choice in frogs shows that his approach to understand mate choice "through the brain of the female"—a combination of neurobiology, evolutionary approach and animal behavior experiments—made him emphasize female agency and contribute to recognizing females as active.

## Fish Research

Sexual selection in fish initially also focused on competition between large males and female choice of large and colorful males. However, some early studies also concerned female choice for male egg guarding and male choice of female size. Subsequent research has nuanced the picture even more, females may compete for mates, prefer to mate with collaborative smaller males rather than large fighters, and the evolution of female color has been explored. In comparison to other vertebrates, male care is relatively common in fishes and research on pipefish and seahorses—in which males brood their offspring in pouches or glued onto their stomach—have been conducted to a large degree by women, of which some are self-declared feminists. These studies demonstrated competition for mates in females as well as in males, thereby contributing to perceptions of females as active and competitive. A group of feminist fish researchers have especially foregrounded the role of females in sexual selection and animal behavior generally.

## Spider Research

Spider females are larger and more sedentary than males, as well as easier to study than males—therefore there has always been a strong focus on females in spider research. However, sexual selection research on spiders has largely focused on sexual cannibalism, why males are so small, female choice and sperm competition—primarily in species that are exceptional by spider standards, by using visual and acoustic signals.[39] Marie Herberstein's situated knowleges of sex and sexual selection are formed by her natural history interest, her gender-reflective perspectives on the research focus on explaining small male size, and the specific biology of

[39] Huber, 2005.

spiders—females are large and in control, sometimes even of their mate's immediate survival. Yet, even when discovering cryptic female choice in her study on post-copulatory sexual selection, sperm competition was the default assumption[40] according to standard procedure in post-copulatory sexual selection experiments.

## Not a Uniform Shift

As the diverse developments of different subfields show, the female turn was not a uniform shift—perceptions about females have shifted at different times and to a greater or lesser degree. The primatologists I interviewed unanimously agreed that there has been a female turn, bird researchers agree to the shift in perceptions of female multiple mating, but consider the perception of passive female birds to still prevail—and so on. There are many different views among evolutionary biologists and there is no clear consensus, as science historian Evelyn Fox Keller aptly describes it—science is a polyphonic chorus.[41] Furthermore, the subfields are not closed scientific communities, they all belong to the larger scientific community of evolutionary biologists and more or less know each other, personally or about each others' research. They sometimes cite each other, in the interviews they often refer to researchers in other subfields, and they also teach about evolutionary biology beyond their own subfield. In addition, as the analysis shows, production of ignorance about females continues.

## The Study Species' Influences

One important conclusion that emerges from this analysis is the large influence that the study species have on "their" researchers' situated knowledges of sex and sexual selection. The active role of the research object is an often under-appreciated aspect of Donna Haraway's concept of situated knowledges. Haraway underlines the agency of nature, she

[40] Elgar et al., 2000.

[41] Keller, 1983: 174.

writes that "the world encountered in knowledge projects is an active entity"[42] and emphasizes acknowledging the unsettling agency of the world as a "witty agent"[43] or a "trickster."[44] In her analysis of primatology, she remarks that "From the point of view of the biologist's purposes, the animals resist, enable, disrupt, engage, constrain, and display."[45] Scientific practice is a story-telling practice in which both the scientist and the organisms are actors.[46] In a similar line, science analyst Bruno Latour has underscored the agency of non-human actors in the knowledge-making process of science.[47] This analysis concretizes how that agency takes form, that it is the study organism which through (in interaction with) the researcher's particular observation/experiment/methodology forms the researcher's situated knowledges.

Furthermore, a recent study by insect and spider researchers Pietro Pollo and Michael Kasumovic at the University of New South Wales in Australia corroborates the findings that researchers' experiences of their study species influence their understanding of sex and sexuality. They investigated the role of biological researchers' country and experiences on their compliance with the typical Darwin-Bateman paradigm's presentation of "sex-roles"[48] (that males compete and are indiscriminate in mate choice while females are choosy) correctly represents patterns among non-human animals.[49] Researchers with experience of working with sexual selection mechanisms contrary to those predicted in typical "sex-roles", for example female-female competition and male choice, were less likely to agree that the typical presentation of "sex-roles" correctly represents patterns in nature. While researchers with experiences of patterns conforming to typical "sex roles", such as female choice, to a larger degree agreed to the typical presentation. Furthermore, they found that researchers spending most of their life in less gender-equal countries were more

[42] Haraway, 1988: 593.

[43] Ibid.: 593.

[44] Trickster is a configuration in Southwest native American myths.

[45] Haraway, 1989: 310.

[46] Ibid.: 4–5.

[47] Latour, 2005.

[48] But see criticism of the sex role concept in Ah-King & Ahnesjö, 2013.

[49] Pollo & Kasumovic, 2022.

likely to agree to the typical view. Both of these findings corroborate my findings in the history of the female turn, namely that researchers' situated knowledges are formed by their experiences—their social experiences as well as their investigations of specific study species.

The history of the female turn demonstrates the influence of feminist critique and gender-reflective approaches in unmaking ignorance about females in science, but also how specific species, cultural perspectives and particular methodologies influence how researchers understand sex and sexual selection. In the subfields studying primates, birds and fishes, feminist scientists have used gender-reflective approaches to challenge perceptions about females, but in other subfields mostly other influences—such as new technologies, studying particular species or interdisciplinary methodology—led to the development toward new understandings of females. This history of situated knowledges in sexual selection underlines how diversity in perspectives, among researchers as well as in study species, provide "many heterogeneous accounts of the world,"[50] and opens possibilities of new knowledge because, as science philosopher Helen Longino emphasizes, knowledge is constructed by "an interactive dialogic community."[51] Haraway underscores that "situated knowledges are about communities, not about isolated individuals,"[52] about "the joining of partial views" in order to provide "better accounts of the world", in sexual selection research, the diverse accounts from different perspectives and about different study species give a more comprehensive understanding of how sexual selection works both in females and males.

## Mechanisms of Ignorance Production in Sexual Selection

The analysis of the female turn shows many different ways in which ignorance about females has been produced. Here, I summarize the different mechanisms of ignorance production that I have identified.

[50] Haraway, 1988: 598.
[51] Longino, 2006: 272.
[52] Haraway, 1988: 590.

## A Repeated Pattern of Male Precedence

Throughout the history of sexual selection, there is a frequent pattern of male precedence, a kind of androcentrism.[53] Male precedence in studying male ornamentation/traits is partly due to the common biological occurrence of extravagant traits in males; however, assumptions of coy, passive and invariant females have also influenced research. For example, a recent study demonstrates widespread female birdsong,[54] contradicting prevailing ideas about birdsong as an exclusively male sexually selected trait, and indicates that there has been a theoretical over-focus on sexual selection on male characteristics to the detriment of understanding female characteristics. This specific study on birdsong also illustrates a geographical bias, since male song is common in the northern hemisphere (where most bird studies have been conducted), while both female and male song is widespread in the tropics and in Australia. Furthermore, there was male precedence in the interpretations of female multiple mating in birds when DNA-analysis revealed surprising levels of "extra-pair" paternity, they were initially explained by male strategies and later studies forwarded active female sexual strategies.

Likewise, the history of postcopulatory sexual selection shows that the male-focused sperm competition hypothesis[55] was suggested first and the equivalent female-focused hypothesis of cryptic female choice was suggested later, another example of male precedence. Sperm competition came to dominate, and still dominates, post-copulatory sexual selection studies. There are practical reasons for this male bias—female internal organs are much harder to study than male external organs or sperm morphology, and equivalent methodologies have not been developed to study the female processes to the same extent as for sperm competition, but there are also theoretical reasons—androcentrism and ignorance of females' potentially active roles. Thus, although female multiple mating

[53] The following section about male precedence is a revised version republished from Ah-King 2022 under the Creative Commons CC BY 4.0 license. Please link to the Creative Commons license: https://creativecommons.org/licenses/by/4.0/.

[54] Odom et al., 2014.

[55] Parker, 1970.

was acknowledged early in insect studies, female agency in post-copulatory sexual selection was recognized relatively late, and although theoretically endorsed, it is still often not studied.

Similarly, genital evolution research has been primarily focused on males and later investigated the evolution of female genitals.[56] The external male genitals are more easily studied than the often internal female genitals (in contrast, female spiders have external genitals[57]). As male genitals are extremely diverse, taxonomists have used them as characteristics for distinguishing species and this wealth of collected information on male genitals was the basis for studies of genital evolution.[58] In addition, female genitals are often soft tissue and may therefore be practically more difficult to study, and they were also assumed to be largely invariant. Subsequent research has found female genitals to evolve intricate structures that enable females to exert cryptic female choice, for example in waterfowl.[59]

The relatively new field of sexual conflict[60] provides another example of male precedence. It has focused especially on male reproductive traits which incur a cost to female fitness during or after copulation[61] and research on the equivalent female adaptations has lagged behind.[62]

This repeated male precedence shows that ignorance about females has been produced throughout the history of sexual selection, even when there were no biological reasons for focusing on males first. The theoretical male-centered focus has blocked, delayed and distorted knowledge about females.

## Loss of Acquired Knowledge

Androcentrism has repeatedly led researchers to overlook available scientific information about active females—thus producing ignorance

[56] Ah-King et al., 2014, Ah-King 2022.

[57] Ibid.: 6.

[58] Eberhard, 1985.

[59] Brennan et al., 2007.

[60] Sexual conflict arises when evolutionary interests over reproduction differ between the sexes.

[61] Arnqvist & Rowe, 1995; Rowe & Day, 2006.

[62] Plakke et al., 2015.

through the loss of acquired knowledge. Early studies describing females as active have often been overlooked—for example in birds and primates. Hypotheses that present females as active—cryptic female choice and the female fertility assurance hypothesis (suggesting that females mate with multiple males to assure fertilization of eggs)—have at least initially been resisted by researchers supporting competing hypotheses of sperm competition and sexual conflict—both of which are predominantly focused on males. When Geoff Parker formulated his theory of male post-copulatory sexual selection, that is sperm competition, he cited a study showing that females affect sperm viability which might have led him to suggest the equivalent female hypothesis, but he did not. Similarly, early studies demonstrating female multiple mating were overlooked in research on birds, snakes and lizards. This is not surprising given the male-centered focus of sexual selection theory.

Another example of loss of acquired knowledge is when anthropologist Donald Symons argued that there was only evidence of female orgasms in humans—and discounted laboratory and zoo observations of non-human animals as unnatural. This rejection of the evidence of female embodied orgasms from laboratory and zoo observations, science philosopher Nancy Tuana describes as constructed through the specific exclusion of evidence. In order to support his story of the evolution of human heterosexual pair-bond, female orgasm needed to be uniquely human.[63]

## Undermining Authority of Certain Knowers

Feminist philosopher Carla Fehr calls attention to the importance of considering "how one decides who counts as an authoritative knower" and that undermining the authority of certain knowers can hinder the production of knowledge.[64] One example is the nineteenth-century Darwinian feminists who criticized the male bias in sexual selection, but since it was at the time when science became professionalized, institutionalized and excluded women, their perspectives were not included in the subsequent production of scientific knowledge. Another example of

[63] Tuana, 2004: 222.

[64] Fehr, 2007: 105.

producing ignorance through disregarding certain knowers was the skepticism toward findings by people working at zoos, that certain lizards and snakes can opt to reproduce without fertilization. Similarly, not long ago, western primatologists seldom cited Japanese primatologists, due to the skepticism of what was considered atheoretical methods and anthropomorphism.[65] Although much of the Japanese primatologists' research has been acknowledged among western primatologists, a cultural divide persists and ignorance production still continues.[66]

## Cultivating Ignorance by Not Citing

Practices of not citing are also a mechanism of ignorance production. Proponents of feminist critique and gender-reflective approaches have experienced this kind of treatment. For example, feminist primatologist Sarah Blaffer Hrdy's book on male and female sexual strategies in langurs was mainly cited for its findings on males.[67] Interestingly, in the influential book *Sexual Selection* (1994) which summarized the field, Malte Andersson, in his section of critique of sexual selection, did not cite feminist critique, but feminist biologists' empirical studies were included in other parts of the book. Oddly, pioneering feminist ornithologist Patricia Gowaty is not mentioned among the 500 leading ornithologists in a historical book on ornithology since Darwin,[68] although elsewhere she is included as one of the leaders in animal behavior.[69] Similarly, leading researchers in post-copulatory sexual selection do not cite clear evidence of cryptic female choice in Simon Griffith and colleagues' Gouldian finch study, thereby cultivating ignorance.

In addition to anecdotal accounts about individual researchers and studies, there are also studies of citation bias. These studies show that in ecology and evolution, studies in the tropics are underrepresented compared to temperate regions, there is taxonomic bias and in sexual

[65] Asquith, 1996: 252ff.
[66] de Waal, 2003.
[67] Sarah Blaffer Hrdy interview 2017.
[68] Birkhead et al., 2014.
[69] Drickamer & Dewsbury, 2010.

selection studies there is citation bias—temperate systems receive higher citation rates.[70] Furthermore, there is citation bias relating to the researchers' nationality,[71] whether the paper supports the tested hypothesis (positive results get more cited), the number of authors (papers with four or more authors were more cited) and the status of the university affiliation.[72]

The citation hierarchy within sexual selection studies is another example of ignorance cultivation through practices of not citing. Birds—as the "paradigmatic" animals of sexual selection studies—are at the top, well read and cited among the researchers working on other animal groups, while bird researchers (as well as mammal researchers) cite within their own animal group to a higher degree than researchers working on other animals.[73] This taxonomic ignorance in sexual selection is generally acknowledged.

This epistemology of ignorance about females (and certain animals) in sexual selection illustrates the different ways in which ignorance has been and continues to be produced. Importantly, the repeated pattern of male precedence—which occurs also when new areas of research are established and there is no logical reason to focus primarily on males—underscores how androcentrism continues to affect sexual selection. In the following sections, I explore continued ignorance production about females in sexual selection.

## Continued Male Bias

Sexual selection theory focuses on how certain males get more mates and offspring than others—through male-male competition or female choice—which emphasizes sexual selection in males, while sexual selection in females has generally been assumed to be lesser or non-existent. Figure 7.1 shows that studies focusing on sexual selection in males far outnumber equivalent studies in females.

---

[70] Culumber et al., 2019.

[71] Paris et al., 1998.

[72] Leimu & Koricheva, 2005.

[73] Taborsky, 2009.

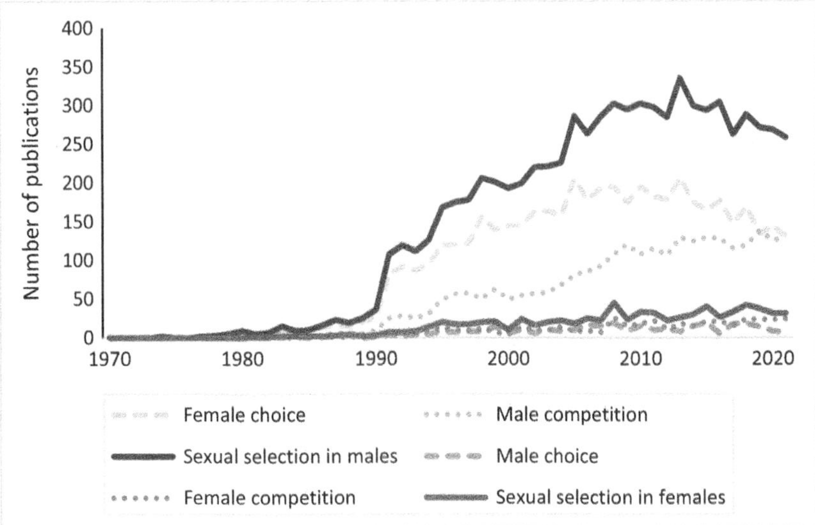


**Fig. 7.1** Number of publications on female choice and male competition, both mechanisms concern variance in male reproductive success, and added together gives the number of publications on sexual selection in males (solid green line)—compared with publications on male choice and female competition, mechanisms which concern the variance in female reproductive success, which added together gives the number of publications on selection in females (solid blue line). The figure shows that studies investigating sexual selection in males outnumber studies of sexual selection in females. Data from Web of Science (June 6, 2022). Reprinted from Ah-King (2022) under the Creative Commons CC BY 4.0 license

## Contrasting Sexual Selection in Males and Females

Figure 7.1 shows the number of publications over time on sexual selection in males versus females. The figure shows studies of four sexual selection mechanisms: female choice, male competition, male choice and female competition. The first two mechanisms concern how males come to differ among themselves in reproductive success, thus sexual selection in males, and the latter two respectively concern sexual selection in females. There is an overwhelming bias toward studies on sexual selection in males. The data is collected from the Web of Science searching "all fields" with the search terms "sexual selection", "female choice", "male competition", "female competition" and "male choice" respectively (June 6, 2022).

## Why Is Sexual Selection in Females Still Understudied?

As I elaborate below, there are a number of current impediments to research on sexual selection in females. It is also important to remember that sex differences are not solely due to sexual selection but also evolve in response to ecological selection pressures.[74] Evolution through differential survival or reproduction may lead to increasing or decreasing sex differences. Evolution is an ongoing process, therefore sex differences are dynamically changing[75] according to complex combinations of several selection pressures—fecundity selection, sexual selection, ecological divergence (adaptations to the physical environment and competition for resources) and (as I explain below) social selection.

### An Alternative Female Turn

In this book, I have described a female turn in evolutionary biology as a turn in perceptions about females. However, an alternative way of characterizing a female turn is in terms of what we know about sexual selection in females—and from this perspective, the female turn has merely started and still has a long way to go. Ornithologist Patricia Gowaty suggests that the main problem in sexual selection is the prevailing assumption that sexual selection in females follows the same pattern as in males. It doesn't pay attention to the ways in which sexual selection in females differ from sexual selection in males.[76] While males do compete for mates and fertilizations, females more often compete for other resources in relation to reproduction, which leads me to deliberate on the definition of sexual selection.

[74] Shine, 1989.
[75] Ah-King & Nylin, 2010.
[76] Altmann, 1997, Hoquet et al., 2020.

## Sexual Selection Definitions

Darwin presented two different definitions of sexual selection, the first in *The Origin of Species* (1859), was a narrow-sense definition,[77] restricted to males: "Sexual Selection … depends, … on a *struggle between the males for possession of the females.*"[78] Later, in *The Descent of Man*, he provided a broad-sense definition,[79] formulated in gender-neutral terms, thus including sexual selection in females as well as competition for other resources than mates: "Sexual selection … depends on *the advantage which certain individuals have over other individuals of the same sex and species, in exclusive relation to reproduction.*"[80]

Presently, the currently prevailing definition is "differences in reproduction that arise from variation among individuals in traits that affect *success in competition over mates and fertilizations.*"[81] Thus, the currently prevailing definition is also a narrow sense definition, which excludes many phenomena that are included in Darwin's broad sense definition, especially concerning competition between females—namely female-female aggression,[82] competition between females for reproductive opportunities to conceive or rear young,[83] and laying eggs in other females' nests.[84]

Patricia Gowaty argues that the issue of the definition of sexual selection is semantic and as long as each study focuses on the specific mechanisms of selection it is not important whether it is called sexual selection.[85] However, semantics and the material are intimately connected, with Haraway's terminology, science is a material-semiotic practice—and categorizations of inside and outside boundaries of sexual selection influence the content of knowledge production, conferences, journals and

[77] Gowaty, 2011.

[78] Darwin, 1859/2006: 103 (my italics).

[79] Gowaty, 2011.

[80] Darwin, 1871/1981, part II: 256 (my italics).

[81] Andersson, 1994: 31 (my italics).

[82] Gowaty, 1985.

[83] Clutton-Brock, 2007.

[84] Gowaty, 1985.

[85] Gowaty, 2011.

funding of sexual selection research. In an attempt to include more female competition in sexual selection, British zoologist Tim Clutton-Brock suggested that we should "return to a broader definition of sexual selection as a process operating through intrasexual [within-sex] competition for reproductive opportunities."[86] However, this suggestion has not been generally accepted in the field,[87] mainly because it would make it even harder to distinguish the boundaries between what is considered to be sexual selection and natural selection.[88]

## Sexual Selection Is Assumed to Be Lesser or Non-existent in Females

According to prevalent assumptions, sexual selection is stronger in males, except in rare cases in which males provide more parental investment than females. Although sexual selection is often stronger in males, evolutionary biologist Zuleyma Tang-Martínez points out (in her review of the current empirical support for Bateman's principles) that the study of female-female competition is still in its infancy. Therefore, it is premature to conclude that male reproductive competition is generally more common or more intense than female reproductive competition.[89] Fish researcher Ingo Schlupp recently published *Male Choice, Female Competition, and Female Ornaments in Sexual Selection* (2021) reviewing the understudied issues of male choice and female-female competition, and a recent meta-study concludes that females generally, across a number of species, increase their reproductive success by mating with multiple males (contrary to Bateman's conclusions).[90] It is telling that both of these studies came out in 2021, one as a published book and the other distributed as a preprint, which suggests that perhaps the assumption of negligible sexual selection in females is about to change.

---

[86] Clutton-Brock, 2007.

[87] See response by Shuker, 2010; Mike Ryan and Darryl Gwynne interviews 2017.

[88] Suzanne Alonzo and Servedio (2019) suggested that researchers pay more attention to the grey zones of sexual selection—the cases in which the definition is ambiguous or controversial.

[89] Tang-Martínez, 2016.

[90] Fromonteil et al., 2021.

## The Overlooked Issue of Social Selection

The issue of social selection has been largely overlooked in evolutionary biology. It was theoretical biologist Mary Jane West Eberhard that provided another way of categorizing competition between females (as well as between males and between the sexes), namely the theoretical framework of social selection—a sub-type of natural selection and itself including the sub-type sexual selection. She defines social selection as "differential reproductive success (ultimately, differential gene replication) due to differential success in social competition, whatever the resource at stake."[91] Thus, social selection includes the many different ways in which social competition and cooperation occur in females and males.[92] A review on female ornamental traits shows that within the same species, the same trait often has different functions for females and males (although there is a continuum of functions between sexual selection and non-sexual social competition in both males and females), but ecological competition is more common in females than in males.[93] Thus, investigating social selection could be one route forward to explore the competition between females (irrespective of the outcome of the sexual selection definition debates).

## Harder to Study Females

Competition between males is often conspicuous, while competition between females often takes more subtle forms. At least among some animals, such as birds, it is harder to study sexual selection in females in practice—following females in the field is much more time-consuming and strenuous (than, e.g., collecting tissue samples and analyzing parentage), and studying female post-copulatory sexual selection inside the female reproductive tract is difficult. The difficulties differ between species, but at least in post-copulatory sexual selection, it seems to be

[91] West-Eberhard, 1983: 158.
[92] West-Eberhard, 2014.
[93] Tobias et al., 2012.

generally harder to study the female side.[94] Evolutionary biologist Caitlin McDonough-Goldstein points out that part of what makes it harder to study the female side is the lack of research priority by the scientific community as well as the lack of established methodology, rather than purely practical obstacles.[95] In spiders, on the other hand—perhaps as an exception—females are to a large degree easier to study than males, and furthermore, they too have external genitals.

Thus, the general theoretical focus on males in sexual selection, the prevalent assumption that sexual selection in females is less important, the prevailing definition of sexual selection and practical difficulties jointly contribute to how sexual selection in females has been and continues to be understudied.[96] Accordingly, many of the same obstacles that historically hindered and slowed knowledge production about females in the field continue to do so.

## Gender and Science

Feminist science studies have shown how cultural assumptions about gender—conceptions about male and female, masculine and feminine as complementary, binary opposites—have influenced science, through their embedding in theory, through language use and metaphors and through the structuring of scientific knowledge. Science historian and philosopher Evelyn Fox Keller analyzed gendered metaphors in science and demonstrated that what has been considered male or masculine has been constructed as opposite to that which is female and feminine—for example, objectivity and subjectivity, reason and emotion, mind and body.[97] Thus, boundaries drawn around gender and science have excluded women, denoted what has been considered feminine as not scientific, and fostered conceptions of science as masculine.[98]

---

[94] Ah-King et al., 2014.

[95] Caitlin McDonough-Goldstein Personal Communication 2021.

[96] The above sections on the sexual selection definition and the following discussion are a revised version from Ah-King 2022, republished under the Creative Commons CC BY 4.0 license. Please link the Creative Commons license: https://creativecommons.org/licenses/by/4.0.

[97] Keller, 1992.

[98] Gieryn, 2011: 23ff.

Accordingly, science historians and philosophers have found connections between the exclusion of women and others and the type of science produced.[99] Therefore, feminist philosophers have criticized the conception of science as objective, value-free and unbiased.[100] Since shifting away from being a male-exclusive domain, science has been improved by the incorporation of a diversity of perspectives and gender-reflective approaches. Thus, contrary to common perceptions of science, mainstream science without the inclusion of feminist critique and gender-reflective perspectives is less objective, and the incorporation of these approaches makes science more objective. Based on the feminist criticism of scientific objectivity, science philosopher Helen Longino developed a theory about validating scientific knowledge called "contextual empiricism", which takes the social context of scientific practice into account. She argues that research constructed by an adequately diverse community with an open critical dialogue and serious consideration of all relevant perspectives is important for enabling progress in scientific knowledge.[101] Thus, she asserts that in order to reach objectivity, science must be open to and engage in "transformative criticism".[102]

Feminist historians of science have been especially interested in the fields of primatology and developmental biology because the influx of women in these particular fields and feminist interventions were associated with the transformation of scientific knowledge—changing images of the role of females in primate societies[103] and the understanding of maternal effects on gene expression[104]—and can thus be seen as victories for contemporary feminists.[105] Evelyn Fox Keller warns that we should not assume that an increase of women in a field would in itself result in shifting paradigms or different ways of doing science—scientific shifts need to be comprehended through "the context of a complex network of interacting social, technical, and cognitive factors."[106]

---

[99] Harding.

[100] Harding, 1986, Haraway, 1988, Nelson, 2017.

[101] Longino, 1990, 2006.

[102] Longino, 1990: 76.

[103] Keller, 2000: 383.

[104] Ibid.: 389.

[105] Ibid.: 382.

[106] Ibid.: 397.

This book details such transitions in the subfields of sexual selection. In sexual selection, feminist critique was important as an incentive for some researchers (especially women), enabling the exposure of gender bias and providing motivation for focusing on female animals in their research. Thus, they exerted epistemological resistance and provided opportunities for, in Nancy Tuana's words, "understanding how to identify, critique and transform ignorance."[107] However, even for biologists who do not identify as feminists, the societal context of the progress by the feminist movement, the entrance of women into science and the sexual revolution changed conceptions of gender in society. In this field (with a prevailing theoretical focus on males), both women and men have developed gender-reflective approaches and investigated ideas focused on females. Many researchers came to their conclusions about active, multiply mating females by their empirical observations of specific animals. Yet, early scattered reports about sexually assertive females in bird and primate research did not change the scientific consensus of perceiving females as passive, because as science philosopher Helen Longino asserts, scientific knowledge "is constructed not by individuals but by an interactive dialogic community."[108] The researchers needed "specific *ways* of seeing"[109] in order to pay attention to active females. As I have described above, important cultural, theoretical and methodological factors combined enabled the scientific community to step-by-step (and at least partly) abandon long-held assumptions about coy, passive and monogamous females. Thus, new combinations of people/politics/science/organisms enabled this intellectual change in evolutionary biology.

Londa Schiebinger argues that in order to improve science, "Scientists need to become aware not just how history shapes the present, but also how what is studied—and what has been neglected—grows out of who is doing the studying, and for what ends."[110] Schiebinger urges for the development of healthy working relationships between those developing gender analyses of science and those doing science. She contends that

---

[107] Tuana, 2006: 1.

[108] Longino, 2006: 272.

[109] Haraway, 1988: 583.

[110] Schiebinger, 1996: 177.

gender awareness needs to be raised within science—through collaborations between gender studies scholars and scientists, through feminist scientists' endeavors, through science students learning about the history of gender in science—and thereby gaining an understanding of the history of women in science as well as the ways in which feminist critique and gender-reflective approaches have changed science.[111]

Thus, a key conclusion is that gender perspectives are important to understand the history of shifting perceptions about females, and also to shed light on what continues to be neglected in sexual selection research. Increased awareness of how gender constructions influence knowledge production may speed scientific progress in this field.

## My Situated Knowledges About the Female Turn

My "specific *ways* of seeing"[112] have formed my situated knowledges about the history of sexual selection. Being brought up in Sweden, with the gender equality ideal, conducting my PhD studies in the male-dominated culture of the Zoology department at Stockholm University, as well as my supervisor, Birgitta Tullberg, the only senior woman at the division of Ecology in 1999, assigning me Sarah Blaffer Hrdy's *The Woman that Never Evolved*[113] as my first reading, cultivated my feminist awareness. So did also my later participation in the Zoology department's gender equality group and seminars, in which we discussed both the Swedish Medical Research Board scandal following Christine Wennerås and Agnes Wold's revelation of sexism and nepotism in peer review,[114] and gender issues in evolutionary biology. My PhD thesis on the evolution of parental care, investigating the environmental and social circumstances of the development of male versus female care of offspring, lay the ground for my later contribution to the growing debate about causes and consequences of variability in "sex roles". Thus, in a way, my thesis did

[111] Schiebinger, 1999: 185f.

[112] Haraway, 1988: 583.

[113] Hrdy, 1981/1999.

[114] Wennerås & Wold, 1997.

include gender issues even if I did not think of it in that way at the time. After my PhD, my interest in gender and evolutionary biology led to a postdoc at the Centre for Gender Research in Uppsala, in which I problematized notions of sex as dichotomous and stable by exploring variation in sex and sexuality among animals,[115] argued for understanding sex as dynamic[116] and conducted gender analyses of evolutionary research.[117] In Uppsala, I also read up on the tradition of feminist biology which I had until then barely been aware of—feminist biologists and analysts of science who pointed out androcentrism in science and the neglect of females in sexual selection—and made contact with Patricia Gowaty, which led to an additional postdoc with her.

These circumstances have formed my understanding of gender bias and the influence of feminist critique in evolutionary biology, the role of the societal and cultural context on biology researchers' knowledges and my recognition of shifting perceptions about females in the field. They also became my incentive for investigating how researchers have theoretized and investigated female animals in sexual selection. With my understanding of all knowledges as situated, it became important to investigate how different researchers' knowledges were situated to understand how and why some came to know about female multiple mating and active females and others did not—how ignorance about females had been produced. Thus, as an "insider-outsider" evolutionary biologist and gender researcher, I understand the ongoing female turn in sexual selection as an effect of the contributing researchers' situated knowledges—formed by their theoretical perspectives, methodologies, specific study animals, personal experiences, feminist insights and gender-reflective approaches—which have shaped knowledge and ignorance production.

Writing the history of a field is also partaking in negotiation of knowledge.[118] By increasing knowledge about how ignorance is produced in evolutionary science, this book hopefully contributes to the development of sexual selection research, as well as to advances in feminist science studies.

---

[115] Ah-King & Nylin, 2010, Ah-King & Ahnesjö, 2013.

[116] Ah-King & Nylin, 2010.

[117] Ah-King et al., 2014.

[118] Haraway, 1989: 126.

## References

Ah-King, M. (2022). The history of sexual selection research provides insights as to why females are still understudied. *Nature Communications, 13*(1), 6976. http://doi.org/10.1038/s41467-022-34770-z

Ah-King, M., & Ahnesjö, I. (2013). The "sex-role" concept: A review and evaluation. *Evolutionary Biology, 40*, 461–470.

Ah-King, M., Barron, A., & Herberstein, M. E. (2014). Genital evolution: Why are females still understudied? *PLoS Biology, 12*(5), e1001851.

Ah-King, M., & Nylin, S. (2010). Sex in an evolutionary perspective: Just another reaction norm. *Evolutionary Biology, 37*, 234–246.

Alonzo, S. H., & Servedio, M. R. (2019). Grey zones of sexual selection: Why is finding a modern definition so hard? *Proceedings of the Royal Society of London Series B, 286*, 20191325. https://doi.org/10.1098/rspb.2019.1325

Altmann, J. (1997). Mate choice and intrasexual reproductive competition: Contributions to reproduction that go beyond acquiring more mates. In P. A. Gowaty (Ed.), *Feminism and evolutionary biology, boundaries, intersections and frontiers* (pp. 320–333). Chapman and Hall.

Andersson, M. (1994). *Sexual selection*. Princeton University Press.

Arnqvist, G., & Rowe, L. (1995). Sexual conflict and arms races between the sexes: A morphological adaptation for control of mating in a female insect. *Proceedings of the Royal Society of London Series B, 261*, 123–127.

Asquith, P. (1996). Japanese science and Western hegemonies. In L. Nader (Ed.), *Naked science: Anthropological inquiry into boundaries, power and knowledge* (pp. 239–256). Routledge.

Birkhead, T. R. (1987). Sperm competition in birds. *Trends in Ecology & Evolution, 2*, 268–272.

Birkhead, T. R., Johnson, S. D., & Nettleship, D. N. (1985). Extra-pair matings and mate guarding in the common murre Uria aalge. *Animal Behaviour, 33*, 608–619.

Birkhead, T., Wimpenny, J., & Montgomerie, R. (2014). *Ten thousand birds: Ornithology since Darwin*. Princeton University Press.

Brennan, P. L. R., Prum, R. O., McCracken, K. G., Sorenson, M. D., Wilson, R. E., & Birkhead, T. R. (2007). Coevolution of male and female genital morphology in waterfowl. *PLoS One, 2*(5), e418. https://doi.org/10.1371/journal.pone.0000418

Burley, N. T., & Parker, P. G. (1998). Emerging themes and questions in the study of avian reproductive tactic. In P. G. Parker & N. T. Burley (Eds.), *Avian reproductive tactics: Female and male perspectives*. The American Ornithologists' Union.

Clutton-Brock, T. H. (2007). Sexual selection in males and females. *Science, 318*(5858), 1882–1885.

Culumber, Z. W., Anaya-Rojas, J. M., Booker, W. W., Hooks, A. P., Lange, E. C., Pluer, B., Ramírez-Bullón, N., & Travis, J. (2019). Widespread biases in ecological and evolutionary studies. *Bioscience, 69*, 631–640.

Darwin, C. (1871/1981). *The descent of man and selection in relation to sex*. Princeton University Press. Reprint of the 1871 edition published by J. Murray.

Darwin, C. (1859/2006). *On the origin of species*. Dover Publications.

de Waal, F. (2003). Silent invasion: Imanishi's primatology and cultural bias in science. *Animal Cognition, 6*, 293–299.

Drickamer, L. C., & Dewsbury, D. A. (Eds.). (2010). *Leaders in animal behavior, second generation*. Cambridge University Press.

Eberhard, W. G. (1985). *Sexual selection and animal genitalia*. Harvard University Press.

Elgar, M. A., Schneider, J. M., & Herberstein, M. H. (2000). Female control of paternity in the sexually cannibalistic spider Argiope keyserlingi. *Proceedings of the Royal Society of London Series B, 267*, 2439–2443.

Fedigan, L. M. (2001). The paradox of feminist primatology: The Goddess' discipline? In A. N. H. Creager, E. Lunbeck, & L. Schiebinger (Eds.), *Feminism in twentieth-century science, technology, and medicine*. University of Chicago Press.

Fehr, C. (2007). Are smart men smarter than smart women? The epistemology of ignorance, women and the production of knowledge. In A. M. May (Ed.), *The 'woman question' and higher education: Perspectives on gender and knowledge production in America*. Edward Elgar Publishing Inc.

Fromonteil, S., Winkler, L., Marie-Orleach, L., & Janicke, T. (2021). Sexual selection in females across the animal tree of life. *bioRxiv*. https://doi.org/10.1101/2021.05.25.445581

Gieryn, T. F. (2011). Boundaries of science. In S. Jasanoff, G. E. Markle, J. C. Peterson, & T. Pinch (Eds.), *Handbook of science and technology studies*. SAGE Publications, Inc.

Glock, H.-J., & Kalhat, J. (2018). Linguistic turn. In *Routledge encyclopedia of philosophy*. Taylor and Francis. https://www.rep.routledge.com/articles/thematic/linguistic-turn/v-1

Gowaty, P. A. (1985). Bluebird belligerence. *Natural History, 85*(6), 8–12.

Gowaty, P. A. (2011). What is sexual selection and the short herstory of female trait variation. *Behavioral Ecology, 22*(6), 1146–1147.

Gwynne, D. T., & Bailey, W. J. (1999). Female-female competition in katydids: Sexual selection for increased sensitivity to a male signal? *Evolution, 53*(2), 546–551.

Hamlin, K. A. (2014). *From Eve to evolution: Darwin, science, and women's rights in Gilded Age America*. The University of Chicago Press.
Hann, H. W. (1937). Life history of the Oven-bird in southern Michigan. *Wilson Bulletin, 49*, 145–237.
Haraway, D. J. (1988). Situated knowledges: The science question in feminism and the privilege of partial perspectives. *Feminist Studies, 14*(3), 575–599.
Haraway, D. J. (1989). *Primate visions: Gender, race and nature in the world of modern science*. Routledge.
Harding, S. (1986). *The science question in feminism: Industrial policy in Europe*. Cornell University Press.
Hatchwell, B. J. (1988). Intraspecific variation in extra-pair copulation and mate defence in common guillemots *Uria Aalge*. *Behaviour, 107*(3–4), 157–185.
Hemmings, C. (2011). *Why stories matter: The political grammar of feminist theory*. Duke University Press.
Hoquet, T., Bridges, W. C., & Gowaty, P. A. (2020). Bateman's data inconsistent with "Bateman's principles". *Ecology and Evolution, 10*, 10325–10342. https://doi.org/10.1002/ece3.6420
Hrdy, S. B. (1981/1999). *The woman that never evolved*. Harvard University Press.
Huber, B. A. (2005). Sexual selection research on spiders: Progress and biases. *Biological Reviews, 80*, 363–385.
Keller, E. F. (1983). *A feeling for the organism: The life and work of Barbara McClintock*. W. H. Freeman.
Keller, E. F. (1992). *Secrets of life, secrets of death: Essays on language, gender and science*. Routledge.
Keller, E. F. (2000). Women, gender, and science: Some parallels between primatology and developmental biology. In S. C. Strum & L. M. Fedigan (Eds.), *Primate encounters* (pp. 382–397). University of Chicago Press.
Latour, B. (2005). *Reassembling the social: An introduction to actor-network-theory*. Oxford University Press.
Leimu, R., & Koricheva, J. (2005). What determines the citation frequency of ecological papers? *Trends in Ecology & Evolution, 20*(1), 28–32.
Longino, H. E. (1990). *Science as social knowledge: Values and objectivity in scientific inquiry*. Princeton University Press.
Longino, Helen E. 2006. Subjects, power, and knowledge: Description and prescription in feminist philosophies of science. In: Keller, Evelyn F. And Longino, Helen E. (eds.) *Feminism and science*, pp. 264–279. : Oxford University Press.
Lönngren, A.-S. (2015). *Following the animal: Power, agency, and human-animal transformations in modern, Northern-European literature*. Cambridge Scholars Publishing.

Malin, Ah-King (2022). The history of sexual selection research provides insights as to why females are still understudied. Nature Communications 13(1) 6976 10.1038/s41467-022-34770-z

McKinney, F., Cheng, K., & Bruggers, D. J. (1984). Sperm competition in apparently monogamous birds. In: Smith, R. L. (ed.) Sperm Competition and the Evolution of Animal Mating Systems. Academic Press Inc.

Milam, E. L. (2010). *Looking for a few good males: Female choice in evolutionary biology*. John Hopkins University Press.

Nelson, L. H. (1993). Who knows? What can they know? And when? *Reason Papers, 18*, 45–56.

Nelson, L. H. (2017). *Biology and feminism: A philosophical introduction*. Cambridge University Press.

Odom, K. J., Hall, M. L., Riebel, K., Omland, K. E., & Langmore, N. E. (2014). Female song is widespread and ancestral in song birds. *Nature Communications, 5*, 3379. https://doi.org/10.1038/ncomms4379

Paris, G., De Leo, G., Menozzi, P., & Gatto, M. (1998). Region-based citation bias in science. *Nature, 396*, 210.

Parker, G. A. (1970). Sperm competition and its evolutionary consequences in the insects. *Biological Reviews, 45*, 525–567.

Plakke, M., Deutsch, A., Meslin, C., Clark, N., & Morehouse, N. (2015). Dynamic digestive physiology of a female reproductive organ in a polyandrous butterfly. *Journal of Experimental Biology, 218*(10), 1548–1555. https://doi.org/10.1242/jeb.118323

Pollo, P., & Kasumovic, M. M. (2022). Let's talk about sex roles: What affects perceptions of sex differences in animal behaviour? *Animal Behaviour, 183*, 1–12.

Proctor, R. N., & Schiebinger, L. (Eds.). (2008). *Agnotology: The making and unmaking of ignorance*. Stanford University Press.

Richards, E. (2017). *Darwin and the making of sexual selection*. The University of Chicago Press.

Richardson, S. S. (2013). *Sex itself: The search for male and female in the human genome*. University of Chicago press.

Rivas, J. A. & Burghardt, G. M. (2005). Snake mating systems, behavior, and evolution: the revisionary implications of recent findings. *Journal of Comparative Psychology, 119*(4), 447–454.

Rowe, L., & Day, T. (2006). Detecting sexual conflict and sexually antagonistic coevolution. *Philosophical Transactions of the Royal Society of London. Series B, Biological Sciences, 361*, 277–285.

Schiebinger, L. (1989). *The mind has no sex? Women in the origins of modern science*. Harvard University Press.

Schiebinger, L. (1996). Gender and natural history. In N. Jardine, J. A. Secord, & E. C. Spary (Eds.), *Cultures of natural history* (pp. 163–177). Cambridge University Press.

Schiebinger, L. (1999). *Has feminism changed science?* Harvard University Press.

Schlupp, I. (2021). *Male choice, female competition, and female ornaments in sexual selection*. Oxford University Press.

Shine, R. (1989). Ecological causes for the evolution of sexual dimorphism: A review of the evidence. *Quarterly Review of Biology, 64*(4), 419–461.

Shuker, D. M. (2010). Sexual selection: Endless forms or tangled bank? *Animal Behaviour, 79*(2010), e11–e17.

Smith, S. M. (1988). Extra-pair copulations in Black-Capped Chickadees—The role of the female. *Behaviour, 107*(1–2), 15–23.

Taborsky, M. (2009). Biased citation practice and taxonomic parochialism. *Ethology, 115*(2), 105–111.

Tang-Martínez, Z. (2016). Rethinking Bateman's principles: Challenging persistent myths of sexually reluctant females and promiscuous males. *The Journal of Sex Research, 53*(4–5), 532–559. https://doi.org/10.1080/00224499.2016.1150938

Thornhill, R. (1983). Cryptic female choice and its implications in the Scorpionfly *Harpobittacus nigriceps*. *American Naturalist, 122*(6), 765–788.

Tobias, J. A., Montgomerie, R., & Lyon, B. E. (2012). The evolution of female ornaments and weaponry: Social selection, sexual selection and ecological competition. *Philosophical Transactions of the Royal Society B, 367*, 2274–2293.

Tuana, N. (2004). Coming to understand: Orgasm and the epistemology of ignorance. *Hypatia, 19*(1), 194–232.

Tuana, N. (2006). The speculum of ignorance: The Women's Health Movement and epistemologies of ignorance. *Hypatia, 21*(3), 1–19.

Vandersommers, D. (2016). The "animal turn" in history. *Perspectives on History*, November 3. https://www.historians.org/publications-and-directories/perspectives-on-history/november-2016/the-animal-turn-in-history

Wennerås, C., & Wold, A. (1997). Nepotism and sexism in peer-review. *Nature, 387*, 341–343.

West-Eberhard, M. J. (1983). Sexual selection, social competition, and speciation. *Quarterly Review of Biolology, 58*, 155–183.

West-Eberhard, M. J. (2014). Darwin's forgotten idea: The social essence of sexual selection. *Neuroscience and Biobehavioral Reviews, 46*, 501–508.

Wilder, G. (2012). From optic to topic. The foreclosure effect of historiographic turns. *The American Historical Review, 117*(3), 723–774.

Wagner, Richard H. (1991). Evidence that female razorbills control extra-pair copulations. *Behaviour, 118*, 157–169.

# Appendix 1

## Selection of Interviewees

I made a strategic choice of interviewees from three overlapping categories: pioneers presenting new theories, hypotheses or empirical evidence, leading researchers in sexual selection and researchers working with explicit gender-reflective approaches perspectives. Geoff Parker and Tim Birkhead's (2013)[1] article about the polyandry revolution pointed out the years 1980-2000 as the time when views on females shifted. I focused on researchers that were active during that time period. After my first interviews, I realized that the study species of a researcher affected their perspectives; there are different communities of evolutionary biologists focusing on specific organisms which make up different sub-fields in evolutionary biology. These sub-fields have different histories and perceptions about females, for example, insect researchers knew about female multiple mating early on due to pest control studies. Therefore, I explicitly added interviewees to increase the diversity of taxonomic study groups—eventually including experts on sexual selection in birds, insects, spiders, frogs, snakes, lizards, fishes and primates. My interviewees were

[1] Parker, G. A., & Birkhead, T. R. (2013). Polyandry: The history of a revolution. *Philosophical Transactions of the Royal Society B-Biological Sciences, 368*, 1471–2970.



M. Ah-King, *The Female Turn*, https://doi.org/10.1007/978-981-19-7161-7

at the time based in the US, UK, Canada, Australia and Sweden. The interviewees are all my seniors and a majority but not all of them are retired or on the point of being retired. I traveled to make face-to-face interviews that lasted about 1,5-2 hours, which often took place in the interviewees' offices or homes. In order to gain background information and prepare targeted questions, I consulted scientific and autobiographical publications written by my interviewees, as well as scientific debates they had been involved in, before conducting the interviews. I also had conversations over e-mail with seven additional researchers.

## List of Interviewees

Malte Andersson, pioneer in sexual selection in birds.
Patricia Adair Gowaty, pioneer in female sexual selection in birds.
Simon Griffith, expert on sexual selection in birds.
Darryl Gwynne, pioneer in sexual selection in insects.
Marie E. Herberstein, expert on sexual selection in spiders.
Sarah Blaffer Hrdy, pioneer on sexual selection in primates.
Charlotta Kvarnemo, evolutionary behavioral ecologist working on fishes.
Geoff Parker, pioneer in sperm competition in insects.
Marion Petrie, expert on sexual selection in birds.
Mike Ryan, pioneer in female choice in frogs and toads.
Richard Shine, expert on sexual selection in snakes, turtles, lizards and frogs.
Leigh Simmons, expert on sexual selection in insects.
Meredith Small, expert on female choice in primates.
Zuleyma Tang-Martínez, rodent expert and feminist critic of the Darwin-Bateman paradigm.
Randy Thornhill, pioneer in cryptic female choice.

## Additional Researchers Contacted by E-mail

Wyatt Anderson, population geneticist conducting early studies on female multiple mating in fruit flies.

William Eberhard, expert on genital evolution and cryptic female choice.

Ambika Kamath, lizard expert.

Bart Kempenaers, expert on sexual selection in birds.

Caitlin McDonough-Goldstein, expert on fruit fly *Drosophila melanogaster* female reproductive biology.

Jesús Rivas, snake expert.

Ingo Schlupp, expert on sexual selection in fishes.

# Index[1]



[1] Note: Page numbers followed by 'n' refer to notes.



M. Ah-King, *The Female Turn*, https://doi.org/10.1007/978-981-19-7161-7

The manufacturer's authorised representative in the EU is Springer Nature Customer Service Centre GmbH, Europaplatz 3, 69115 Heidelberg, Germany. If you have any concerns regarding our products, please contact ProductSafety@springernature.com

Printed and bound by CPI Group (UK) Ltd, Croydon, CR0 4YY

30/09/2026

02225243-0009